Control Basics for Mechatronics

Mechatronics is a mongrel, a crossbreed of classic mechanical engineering, the relatively young pup of computer science, the energetic electrical engineering, the pedigree mathematics and the bloodhound of Control Theory.

All too many courses in control theory consist of a diet of 'Everything you could ever need to know about the Laplace Transform' rather than answering 'What happens when your servomotor saturates?' Topics in this book have been selected to answer the questions that the mechatronics student is most likely to raise.

That does not mean that the mathematical aspects have been left out, far from it. The diet here includes matrices, transforms, eigenvectors, differential equations and even the dreaded z transform. But every effort has been made to relate them to practical experience, to make them digestible. They are there for what they can do, not to support pages of mathematical rigour that defines their origins.

The theme running throughout the book is simulation, with simple JavaScript applications that let you experience the dynamics for yourself. There are examples that involve balancing, such as a bicycle following a line, and a balancing trolley that is similar to a Segway. This can be constructed 'for real', with components purchased from the hobby market.

Control Basics for Mechatronics

John Billingsley

CRC Press
Taylor & Francis Group
Boca Raton London New York

CRC Press is an imprint of the
Taylor & Francis Group, an **informa** business

First edition published 2024
by CRC Press
2385 Executive Center Drive, Suite 320, Boca Raton, FL 33431

and by CRC Press
4 Park Square, Milton Park, Abingdon, Oxon, OX14 4RN

CRC Press is an imprint of Taylor & Francis Group, LLC

ISBN: 9781032425573 (hbk)
ISBN: 9781032425832 (pbk)
ISBN: 9781003363316 (ebk)
ISBN: 9781032526799 (eBook+)

DOI: 10.1201/9781003363316

Typeset in Times
by codeMantra

Access the companion website: https://www.routledge.com/cw/Billingsley

Contents

Preface

Mechatronics is a mongrel, a crossbreed of classic mechanical engineering, the relatively young pup of computer science, the energetic electrical engineering, the pedigree mathematics and the bloodhound of Control Theory.

All too many courses in control theory consist of a diet of 'Everything you could ever need to know about the Laplace Transform' rather than answering 'What happens when your servomotor saturates?' Topics in this book have been selected to answer the questions that the mechatronic student is most likely to raise.

That does not mean that the mathematical aspects have been left out, far from it. The diet here includes matrices, transforms, eigenvectors, differential equations and even the dreaded z transform. But every effort has been made to relate them to practical experience, to make them digestible. They are there for what they can do, not to support pages of mathematical rigour that defines their origins.

The theme running throughout the book is simulation, with simple JavaScript applications that let you experience the dynamics for yourself. There are examples that involve balancing, such as a bicycle following a line, and a balancing trolley that is similar to a Segway. This can be constructed 'for real', with components purchased from the hobby market.

Logical Predictive Control updates my PhD research topic of long ago, originally demonstrated at IFAC 66 in London's Festival Hall. My wife Rosalind drew some magnificent posters to illustrate the principle, and they are included here.

I can only hope that you will have as much fun reading the book and controlling the simulations, as I have had writing it.

If you are reading a printed version of the book, you will have to find the simulations on the web. Navigate your browser to:

www.routledge.com/cw/Billingsley

There you will see a landing page that groups the links to simulations by chapter.

You can also find a backup page at **www.esscont.com/sim**, where there might be future updates and corrections.

HOW TO ACCESS AND USE THE SIMULATIONS

You find a link to the simulations and other material for this book at

www.routledge.com/cw/Billingsley

Click the link for this book, and then select the 'sim' tab. Here you will find links, chapter by chapter, to each of the simulations.

But by using the 'camera' application that is already on your laptop, there is an easy way to save typing. The camera can read the printed QR code here.

A link will be copied into the buffer and it can be pasted into your browser. Some QR readers will let you launch the link directly.

For Windows users, to launch the Camera app, just type 'camera' into the search box at the bottom of your screen. When it launches, you will see a bar on the right:

Select the barcode option by clicking the indicated icon.

Now show the QR code to the camera and click the icon.

This will give you access to the simulations whenever you are online.

You can download the simulations for offline use at

https://routledgetextbooks.com/textbooks/9781032425573/sim/sim.zip

Once again, you can save a lot of typing by using this QR code:

Unzip the folder to any convenient place on your drive, then look for the file sim.htm inside it. Clicking it will launch a browser page in which you can insert the name of any of the simulations.

John Billingsley
2023.

MATLAB® is a registered trademark of The Math Works, Inc. For product information, please contact:
The Math Works, Inc.
3 Apple Hill Drive
Natick, MA 01760-2098
Tel: 508-647-7000
Fax: 508-647-7001
E-mail: info@mathworks.com
Web: http://www.mathworks.com

Author

John Billingsley is Professor of Mechatronic Engineering at the University of Southern Queensland, Australia.

1 Why Do You Need Control Theory?

ABSTRACT

Control theory is not just about gadgets. It underpins the behaviour of all dynamic systems, from a nation's economy to the response of a plant to sunshine. It is the study of time itself. But here those aspects have been singled out that relate to the control of devices, mechatronics.

Differential equations govern the way that one property can determine the rate of change of another. A computer can be set up to mimic those equations, to simulate the way that a system will behave.

Control theory is not just about gadgets. It underpins the behaviour of all dynamic systems, from a nation's economy to the winding of a climbing honeysuckle plant. It is the study of time itself. But in this book the aspects have been singled out that relate to mechatronics.

The world is now full of gadgets that owe their operation to embedded microcontrollers. The topic with that rather cumbersome name of 'mechatronics' embraces their mechanical construction, the electronics for sensors of what is happening, more electronics for applying drives to motors and such, plus the software needs of the microcontrollers themselves.

But that is not all. The software must have a strategy for weighing up the evidence of the sensors, to determine what commands should be sent to the outputs. The designer must have an understanding of control theory, the essence of this book.

Over half a century ago, it was realised that there was a more direct way to look at dynamic systems than the *frequency domain*, with its dependence on poles, zeroes and complex numbers. This was the *state space* approach. It is the magic key to the construction of simulations that will let us visualise the effect of any control strategy that we are considering applying. Of course, those transforms do have their uses, but they run into problems when your system has nonlinearities such as drive constraints.

But do not be dismissive of the mathematics that can help us to understand the concepts that underpin a controller's effects. When you write up your control project, the power of mathematical terminology can lift a report that is about simple pragmatic control, to the status of a journal paper.

1.1 CONTROL IS NOT JUST ABOUT ALGORITHMS

We can find early applications of control from long before the age of 'technology'. To flush a toilet, it was once necessary to tip a bucket of water into the pan – and then walk to the pump to refill the bucket. With a piped water supply, a tap could be turned to fill a cistern – but you still had to remember to turn the tap on and off.

DOI: 10.1201/9781003363316-1

Today everyone expects a float inside the cistern to do that for you automatically – you can flush and forget.

The technology that turned a windmill to face the wind needed more ingenuity. These were not the massive wind turbines of today, but the traditional windmills for which Holland is so famous. They were too big and heavy to be rotated by a simple weathervane, so when the millers tired of lugging them round by hand, they added a small secondary fan to do the job. This was mounted at right angles to the main rotor, to catch any crosswind. With gearing it could slowly crank the whole mill around, in the necessary sense to make it face the wind.

Today we can easily simulate either of these systems, but it is most unlikely that any mathematical analysis was used in their original design. The strategies grew out of a simple understanding of the way that the systems would behave.

Take a rest from reading and try a simple experiment. You will need two garden canes, about a metre long, and four tapered disposable drinking mugs. Tape the cups together in pairs, in one case mouth to mouth, in the other by joining the narrower closed ends. You now have two rollers, one bulging in the middle, the other with a narrow waist as in Figure 1.1.

Prop up one end of the canes, so that side by side they make a sort of inclined railway track. In turn, put one of your rollers at the top of the track and watch it roll down. Which one makes it to the bottom without falling off the side?

This is the technique used to keep a railway carriage on the rails. Unlike a toy train set, the flanges on the wheels should only ever touch the rails in a crisis. The control is actually achieved by tapering the wheels, as shown in Figure 1.2. Each pair of wheels is linked by a solid axle, so that the wheels turn in unison.

To see how this works, suppose that the wheels are displaced to the right. The right-hand wheel now rolls forward on a larger diameter than the left one. The right-hand

FIGURE 1.1   The two rollers made from paper cups.

FIGURE 1.2   A pair of railway wheels.

wheel travels a little faster than the left one and the axle turns to the left. Soon it is rolling to the left and the error is corrected. But as we will soon see, the story is more complicated than that. As just described, the axle would 'shimmy', oscillating from side to side. In practice, axles are mounted in pairs to form a 'bogey'. The result is a control system that behaves as needed without a trace of electronics.

The moral is that stability of a mechatronic system can often be aided by ingenious mechanical design. But for mechatronic control we will be concerned with feedback. We will use sensors to measure what the system is doing now and apply control to actuators to make it do what we want it to. This can lead to stability problems, so a large body of theory has been built up for linear systems. There is more about this at the end of the chapter.

Unfortunately, few real systems are truly linear. Motors have limits on how hard they can be driven, for a start. If a passenger aircraft banks at more than an angle of thirty degrees, there will probably be complaints if not screams from the passengers. Methods are needed for *simulating* such systems, for finding how they respond as a function of time.

1.2 THE ORIGINS OF SIMULATION

The heart of a simulation is the *integrator*. This gives the means to find the response corresponding to a differential equation. If the output of an integrator is x, then its input is dx/dt. By cascading integrators, we can construct a differential equation of virtually any order. But where can we find an integrator?

In the Second World War, bomb-aiming computers used the 'ball-and-plate' integrator. A disk rotated at constant speed. A ball bearing was located between the plate and a roller, being moved from side to side as shown in Figure 1.3. When the ball is held at the centre of the plate, it does not move, so neither does the roller. If it is moved outwards along the roller, it will pick up a rotation proportional to the distance from the centre, so the roller will turn at a proportional speed. We have an integrator!

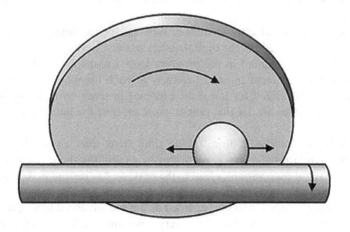

FIGURE 1.3 Ball-and-plate integrator.

But for precision simulation, a 'no-moving-parts' electronic system was needed. A capacitor is an electronic component which builds up a voltage proportional to the integral of the current that has flowed through it. By using a capacitor to apply feedback around an amplifier, we have an integrator.

Unfortunately, in the days of valves the amplifiers were not easy to make. The output had to vary to both positive and negative voltages, for a very small change in an input voltage that was near zero. Conventional amplifiers were *AC coupled*, being used for amplifying speech or music. These new amplifiers had to give a constant DC output for a constant input and were annoyingly apt to drift.

But in the early 1960s, the newfangled transistor came to the rescue. By then, both PNP and NPN versions were available, allowing the design of circuits where the output was pulled up or down symmetrically.

Within a few years, the manufacturers had started to make 'chips' with complete circuits on them, and an early example was the *operational amplifier*, just the thing the simulator needs. These have become increasingly more sophisticated, while their price has dropped to a few cents.

Just when perfection was in sight for the analogue computer (or simulator), the digital computer moved in as a rival. Rather than having to patch circuitry together, the control engineer only needs to write a few lines of software to guarantee a simulation with no drift, no uncertainty of gain or time constants, and an output that can produce a plot only limited by the engineer's imagination.

While the followers of the frequency domain methods concern themselves with *transfer functions*, simulation requires the use of *state equations*. You just cannot escape mathematics!

1.3   DISCRETE TIME

Simulation has changed the whole way we view control theory. When analogue integrators were connected to simulate a system, each one defined a first-order equation, with the list of its inputs. Its instantaneous output value could be regarded as a *state variable*, and the whole system has been reduced to a set of first-order *state equations*.

Digital simulation added to the impact. Whereas the amplifier voltages changed continuously, the changes of the digital state variables were stepped in time, each time that they were updated in the program loop. Computer simulation and discrete-time control go hand in hand together. At each iteration of the simulation, new values are calculated for the state variables in terms of their previous values. New input values are set that remain constant over the interval until the next iteration.

We might be cautious at first, taking time steps that are so short that the calculation approximates to integration. But by examining the exact way that the values of one set of state variables lead to the next, we can instead make the interval longer.

Discrete-time theory is usually regarded as a more advanced topic than the frequency domain, but in some respect it is very much simpler. Whereas the frequency domain is filled with complex exponentials, discrete-time solutions just involve powers of a parameter – though this may be a complex number, too.

By way of an example, consider your bank overdraft. If the interest rate causes it to double after m months, then after further m months it will double again. After n periods of m months, it will have been multiplied by 2^n. We have a simple solution for calculating its values at these discrete intervals of time.

To calculate the response of a system and to assess the effect of discrete-time feedback, a useful tool is the *z-transform*. This is usually explained in terms of the *Laplace transform*, but its concept is much simpler.

In simulating an integrator, when we calculate the new value of a state variable x from its previous value and the input u, we might have a line of code of the form

```
x = a*x + b*u
```

Of course, this is not an equation. The x on the left is the new value while that on the right is the old value. But we can turn it into an equation by introducing an *operator* that means *next*. We denote this operator as z.

So now

$$z\,x = a\,x + b\,u$$

or

$$x = \frac{bu}{z-a}$$

In later chapters all the mysteries will be revealed, but before that we will explore the more conventional approaches.

You might already have noticed that I prefer to use the mathematician's "we" rather than the more cumbersome passive. Please imagine that we are sitting shoulder to shoulder, together pondering the abstruse equations that we must inevitably deal with.

1.4   THE CONCEPT OF FEEDBACK

For your mechatronic system, you are likely to wish to use large values of gain, for example, to provide a large restoring force if a positioning system is disturbed from its target. The problems of high gain were addressed by the early developers of control theory.

When the early transatlantic cables were laid, amplifiers had to be submerged in mid ocean. It was important to match their 'gain' or amplification factor to the loss of the cable between repeaters. Unfortunately, the thermionic valves used in the amplifiers could vary greatly in their individual gains and that gain would change with time. The concept of feedback came to the rescue. A proportion of the output signal was subtracted from the input. So how does this help?

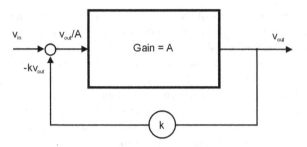

Effect of feedback on gain.

Suppose that the gain of the valve stage is A. Then the input voltage to this stage must be $1/A$ times the output voltage. Now let us also subtract k times the output from the overall input. This input must now be greater by kv_{out}. So the input is given by

$$v_{in} = (1/A + k)\, v_{out}$$

and the gain is given by

$$v_{out}/v_{in} = \frac{1}{k + 1/A}$$

$$= \frac{1/k}{1 + 1/Ak}$$

What does this mean? If A is huge, the gain of the amplifier will be $1/k$. But when A is merely 'big', the gain fails short of expectations by denominator of $1 + 1/(Ak)$. We have exchanged a large 'open loop' gain for a smaller one of a much more certain value. The greater the value of the 'loop gain' Ak, the smaller is the uncertainty.

But feedback is not without its problems. Our desire to make the loop gain very large hits the problem that the output does not change instantaneously with the input. All too often a *phase shift* will impose a limit on the loop gain that we can apply before instability occurs. Just like a badly adjusted public-address microphone, the system will start to 'squeal'.

Over the years, the electronic engineers built up a large body of experience concerning the analysis and adjustment of linear feedback systems. To test the gain of an amplifier, a small sinusoidal 'whistle' from an *oscillator* was applied to the input. A variable *attenuator* could reduce the size of an oscillator's one-volt signal by a factor of, say, a hundred. If the output was then found to be restored to one volt, the *gain* was seen to be one hundred. (As the amplifier said to the attenuator, "Your loss is my gain." I apologise for the pun!)

As the frequency of the oscillator was varied, the gain of the amplifier was seen to change. At high frequencies, it would *roll off* at a rate measured in *decibels per octave* – the oscillators had musical origins and levels were related to 'loudness'.

Some formal theory was needed to validate the rules of thumb that surrounded these plots of gain against frequency. The electronic engineers based their analysis on complex numbers. Soon they had embroidered their methods with Laplace transforms and a wealth of arcane graphical methods, Bode diagrams, Nyquist diagrams, Nicholls charts and root locus to name but a few. Not surprisingly, this approach was termed the *frequency domain*.

When the control engineers were faced with problems like simple position control or the design of autopilots, they had similar reasons for desiring large loop gains. They hit stability problems in just the same way. So they 'borrowed' the frequency-domain theory lock, stock and barrel.

For the design of mechatronic control systems, many of those techniques are not really necessary. But some of them can be extremely useful.

2 Modelling Time

ABSTRACT

After the generalities of chapter one, we see how those differential equations can be put into action. We start to explore the ways that a computer program can be made to dance to their tune, to show us a simulation of "What would happen if?"

2.1 INTRODUCTION

In every control problem, time is involved in some way. It might appear in an obvious way, relating the height at each instant of a spacecraft, in a more subtle way as a list of readings taken once per week, or unexpectedly as a system goes unstable.

Whether the values are logged or not, the involvement of time is through differential or difference equations, linking the system behaviour from one moment to the next. This is best seen with an example.

2.2 A SIMPLE SYSTEM

Figure 2.1 shows a cup of coffee that has just been made. It is rather too hot at the moment, at 80°C. If left for some hours it would cool down to room temperature at 20°C, but just how fast is it going to cool right now, and when will it be at 60°C?

The rate of fall in temperature will be proportional to the rate of loss of heat. It is a reasonable assumption that the rate of loss of heat is proportional to the temperature above ambient, so we see that if we write T for temperature, we have a differential equation

$$\frac{dT}{dt} = k\left(T - T_{ambient}\right).$$

Here k is negative. If we can determine the value of the constant k, perhaps by a simple experiment, then the equation can be solved for any particular initial temperature – the form of the solution comes later.

Equations of this sort apply to a vast range of situations. A rainwater butt has a small leak at the bottom as shown in Figure 2.2. Suppose that the rate of leakage is proportional to the depth, H, so that:

$$\frac{dH}{dt} = -kH$$

The water will leak out until eventually the butt is empty.

DOI: 10.1201/9781003363316-2

9

FIGURE 2.1   A cooling cup of coffee.

FIGURE 2.2   A leaking water butt.

But suppose now that there is a steady flow *into* the butt, sufficient to raise the level (without leak) at a speed u. Then the equation becomes:

$$\frac{dH}{dt} = -kH + u$$

Exercise 2.1

Show that the level will eventually settle at a value of $H = u/k$. Now if we really want to know the depth as a function of time, a mathematical formula can be found for the solution. But let us try another approach first, simulation.

2.3 SIMULATION

With a very little effort, we can construct a computer program that will imitate the behaviour of the water level.

If the depth right now is H, then in a short time dt, the depth will have changed by the rate of change multiplied by the interval

$$(-k.H + u)dt$$

To get the new value of H, we add this to the old value. In programming terms, we can write:

```
H = H + (-k*H + u)*dt;
```

Although it might look like an equation, this is an *assignment statement* that gives a new value to H each time a program loop executes it.

We can add another line to update the time as well,

```
t = t + dt;
```

To make a simulation, we must wrap this in a loop.

```
while (t < tmax) {
   H = H + (-k*H + u)*dt;
   t = t + dt;
}
```

Now the code has to be 'topped' to set initial values for H, t and u, and it will calculate H until t reaches the time *tmax*. But although the computer might 'know the answer', we have not yet added any output statement to let it tell us.

Although a 'print' statement would reveal the answer as a list of numbers, we would really prefer to see a graph. We would also like to be able to change the input as the simulation goes on. So what computing environment can we use?

2.4   CHOOSING A COMPUTING PLATFORM

In the early years, the 'language of choice' would have been something like Quick Basic, Visual C or MATLAB®. Java was popular until security concerns made browsers rule it out. But many languages have in turn become obsolete, or require serious updating. So we need an environment that is likely to endure several generations of software updates by the system vendors.

Internet browsing has now become a dominant application. Every browser now supports a scripting language, usually employed for animating menu items and handling other housekeeping. The *JavaScript* language is very much like *C* in appearance.

Under the umbrella title HTML5, powerful graphics facilities have been released. One of these was *Canvas*, which not only provided a surface to draw on, but gave a toolbox that included transformations, including three-dimensional ones.

As an added bonus, JavaScript contains a command 'eval' that enables you to type your code as text in a box on the web page, and then see it executed in real time. However, eval is deprecated by the 'lords of the internet' and its use is being increasingly restricted.

A villain could insert 'malware' to run on your computer – I hope that you are not such a villain, but take care! Enjoy its use while you can, before it gets phased out.

But there's more.

By putting control buttons on the web page, you can interact with the simulation in real time, in a way that would be difficult with any expensive proprietary package.

The next chapter will outline the use of HTML5 for simulating the examples that we have met so far. We can start with an even simpler simulation, something falling under gravity.

This will have two state variables, the height h and the vertical velocity v. So after each interval dt, h will have $g * dt$ added to it, where of course g is negative.

$$v = v + g * dt$$

The new value for h will be

$$h = h + v * dt + \tfrac{1}{2} g * dt^2$$

If we make dt small enough, we can ignore that second term and simply use the code

```
v = v + g*dt;
h = h + v*dt;
```

3 A Simulation Environment

ABSTRACT

Before we look at any particular problem, we must consider the building of the simulations themselves. Any of a dozen computer languages could be used, but we opt for one that is contained in every browser.

Simulating the equations is one thing, but showing them to the user is another. We look at ways of turning the simulation into an informative moving picture. We can even interact in real time with such a simulation. A Segway simulation is paraded 'just for show', where all the hard work of controlling it has been done for you. But do not be disappointed. A tricky bicycle problem awaits you later.

3.1 JOLLIES

The acronym stands for "Javascript On Line Learning Interactive Environment for Simulation."

Visitors to the website www.jollies.com might be surprised to find an outline, written long ago, of how to exploit JavaScript for simulations. The code can be modified by on-screen editing inside the browser window, without downloading any application. However, that ability might change, as browsers start to restrict the use of 'eval'.

Serious simulations will of course require something more sophisticated, but Jollies puts a text panel below the image, containing the essential code that the student must think about. The 'housekeeping' is hidden from view, while for now the code in the window can be edited to change any feedback strategy or system parameters.

There are two essential features. The first is that 'eval' command that causes the code in a text box to be executed. The second is the 'setTimeout' function that calls a specified routine after a number of milliseconds have elapsed. If that routine contains the same 'eval' instruction, the code will be executed over and over again. Thereby hangs our simulation.

3.2 MORE ON GRAPHICS

The opportunities presented by HTML5 are boundless, but we should start with something simple.

The HTML markup language is simple text that deals with the positioning and style of items to display. These items all have *properties*, whether they are an image, a table, a canvas or a simple piece of text.

DOI: 10.1201/9781003363316-3

13

Those properties include *style.top* and *style.left*, so by changing these, the item can be displayed anywhere on the page. A falling object can be simulated by simple code. The 'engine room' of this code is:

```
<script>
  h=0;
  v=0;
  g=10;
  dt=.01;
function step() {
  v=v+g*dt;
  h=h+v*dt;
  leaf.style.top=h+"px";
  if(h<400){setTimeout('step();',10);}
}
</script>
</head>
<body><p id="leaftext"> Falling Leaf </p>
<script>
  leaf = document.getElementById('leaftext');
  leaf.style.position='absolute';
  leaf.style.top="0px";
  step();
</script>
</body>
```

As you see, it is a simple text file, littered with 'tags'. These tend to come in pairs, with a closing tag starting with a 'slash' </ >.

The first tag here opens the 'head' section where things are prepared for the display, such as styles, variables and any routines. JavaScript is enclosed within <script> and </script> tags.

After the head is closed, the <body> tag starts to list the contents of anything that is to be shown on the page, in this case a text line that includes the word 'Falling Leaf'.

An extra <script> comes after that line, to attach a style handle to it. This could not be done in the head until the line had been introduced. Then step() is called to start the simulation running.

Finally, everything is tidied up with </body> and </html>.

You can copy the text above and paste it into a Notepad file. When you save it and change the extension to.htm, double-clicking on it should cause it to run.

Exercise 3.1

Click on the link to **falling** to see it run. Then right click in the page to make sure that the source is really the same as the code listed here.

But it is not! So that the code can be included in an EPUB, it must be topped with some housekeeping to make it conform to the standards of XML. But the main part is exactly the same.

3.3 MORE CHOICES

Of course, a simulation will not want to stop with simply moving some text around. The text line can be replaced with an image. Replace the line

```
<p id="leaftext"> Falling Leaf </p>
```

with

```
<img id="ball" src="bred.gif"> </img>
```

Exercise 3.2

Visit **fallingball** where you will see a falling ball. Here bred.gif is an image of a red ball in the same folder.

The code in this example has been stripped down to a bare minimum. As we get more serious, we can take advantage of many more features of HTML. We can include 'buttons' where a click will cause some code to be executed. We can include text boxes into which new parameter values can be typed. We can include a 'textarea' in which code can be edited, before a new run is tried again.

Jollies are web pages in which the gory complications are hidden from view, while the student can concentrate on the actual control task. But a right click and 'view source' will make all those details visible, so that the enterprising student can elaborate simulations to his or her heart's content.

Exercise 3.3

Visit **Segway** where there is a 'moving balls' simulation that deals with an advanced control problem. It shows the versatility of this sort of simulation.

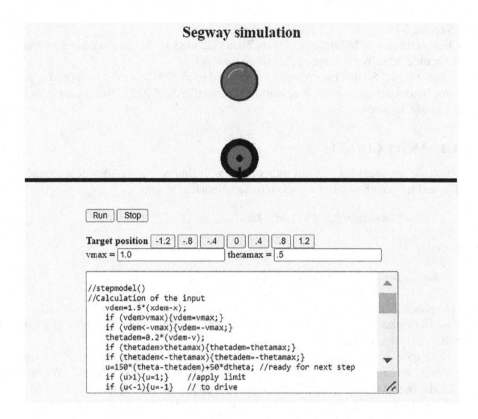

Screenshot of Segway.xhtml.

3.4   DRAWING GRAPHS

Moving images around is a great way to animate a simulation, but publications need graphs. HTML5 provides the canvas that does everything you need.

If you insert the line

```
<canvas id="canvas" width="600" height="400"
style="background:white"></canvas>
```

in the body of your page, you will have a board 600 by 400 pixels to draw on. But before that, you must get its *context*, a kitbag of tools including ways to draw, ways to set line width and colour, and even some powerful transformations.

First, you must put a handle on the canvas with

```
canvas = document.getElementById("canvas");
```

and equip the program with the toolbag

```
ctx = canv.getContext("2d");
```

By way of an example, let us wrap the water butt simulation in a minimum quantity of code to plot a graph in a web page. We start with

```
<html>
<head>
<script>
    var k = .5;
    var dt = .01;       //Edit to try various steplengths
    var t=0;
    var x=0;                //Initial depth
    var u=20;               //Input flow
    function simulate() {
        canvas = document.getElementById("canvas");
        ctx = canv.getContext("2d");
        ctx.strokeStyle = "blue";
        ctx.beginPath();
        ctx.moveTo(50*t, 400-10*x);
//This is the simulation
        while (t<10) {
            x = x + (-k*x + u) * dt;
            t = t + dt
            ctx.lineTo(50*t, 400-10*x);
        }
        ctx.stroke();
    }
</script>
</head>
```

This defines all the variables that we want to use, then sets up the simulate() function to do all the hard work, including setting up the tools.

Now we design a web page with a grey background and the canvas centred

```
<body style="background:gray" onLoad="simulate();">
<canvas id="canvas" width="600" height="400"
style="background:white";></canvas>

</body>
</html>
```

That completes the code. The 'onLoad' inside the body tag does not take effect until everything has been loaded. Then it kicks off the simulation.

Exercise 3.4
Open Notepad and paste in the code above. Save it as a text file with title butt.htm. Do not close the editor.

Open that file with a browser, any should do, and after the page has loaded you will see a graph appear. It should look like Figure 3.1.

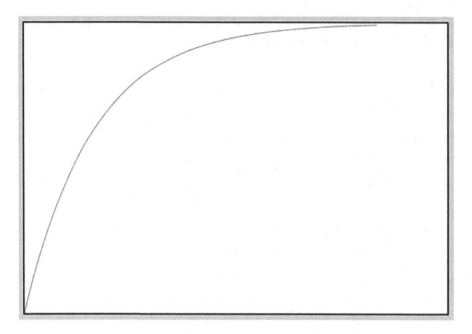

FIGURE 3.1   Screen grab of butt.html.

Exercise 3.5
Now edit the code to give initial conditions of $x = 40$ and $u = 0$. Save the file and reload the browser web page. What do you see?

Exercise 3.6
Edit the file again to set the initial $x = 0$ and $u = 10$. Save it again and reload the web page. What do you see this time?

Exercise 3.7
Edit the file again to set $dt = 1$. What do you see this time?

Exercise 3.8
Now try $dt = 2$.

Exercise 3.9
Finally try $dt = 4$.

3.5   MORE DETAILS OF JOLLIES

It is obvious that there are better ways to try different parameters than to edit them into the source code. By using '*textbox*' and '*textarea*', values can be typed straight into the browser. By adding buttons and slowing down the simulation, inputs can be switched in mid run.

A standard style for the Jollies is as follows:

When the page is first loaded, a button must be clicked to start the simulation. That is to give the reader time to edit the code.

Below the canvas there are two *textareas*, boxes to fill with text. The first of these has the same function as the HTML <head>, setting up variables and defining functions. It runs just once when the 'run' button is clicked.

The second is for the simulation code itself. It is executed by an 'eval' statement in a function in the 'housekeeping code', and ends with a setTimeout 'alarm clock' to call that function to be run again some milliseconds later.

With the simulation environment set up, it will now be possible to concentrate on the control aspects.

The 'engine room' code looks rather messy, with a need to scale variables to give a pixel position in the canvas. To draw a line to (*x*, *y*) we might need code like

```
ctx.lineTo (x0+xsc*x, y0+ysc*y);
```

so a subroutine is made available to do all the housekeeping.

This is contained in an *'include file'* canv.js, together with a routine to initialise the context and scale the canvas.

Now the code to draw the line can simply be

```
LineTo(x, y);
```

The window is scaled using something like

```
ScaleWindow(-1,-5,tmax, 40);
```

to put (–1, –5) in the lower left corner and (tmax, 40) at the top right.

All this will work for now, on code that you write yourself or load from a web site like jollies.com. But for code contained in an EPUB, there are some very strict rules that stand in the way of executing text in a window. This is for the very good reason that a hacker might persuade you to enter some malicious code, perhaps fooling you into thinking that it will get rid of some computer virus, when it is actually loading one.

So, although the simulations in this book will show you panels of the code they use, you will not be able to edit them, they are just text copies.

4 Step Length Considerations

ABSTRACT

A digital computer performs its calculation in 'lumps of time', though these can be very small. Nature moves things smoothly and continuously, so we must find ways to help the computer to avoid getting out of step.

As we do this, we start to get our feet wet in some simple solutions, using a little of the calculus that we are trying to keep until later.

4.1 CHOOSING A STEP LENGTH

You should have noticed that the result of the simulation varies if you change the value of *dt*. For small step lengths the change might be negligible, but for values approaching one second, the approximation deteriorates rapidly. The calculation is based on assuming that the rate-of-change remains constant throughout the step interval, but if the interval is not small the errors will start to build up.

A practical solution is to halve the value of *dt* and repeat the simulation. If no change can be seen, the interval is short enough. Another strategy might be to use a much more complicated integration process, such as Runge–Kutta, but it is preferable to keep the code as simple as possible.

In the exercises at the end of chapter three, you should have seen that if the step length is too long, not only does the simulation deteriorate, it can become unstable.

Here is another simulation on which it is much easier to try out different step lengths. The code has been embellished with a text box to enter the step size and run again.

DOI: 10.1201/9781003363316-4

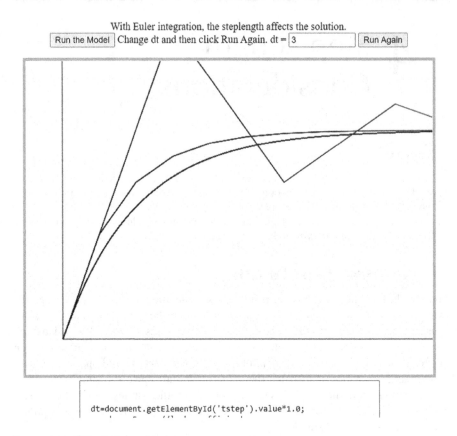

With Euler integration, the steplength affects the solution.

| Run the Model | Change dt and then click Run Again. dt = 3 | Run Again |

```
dt=document.getElementById('tstep').value*1.0;
```

Screenshot of Step1-Euler.xhtml.

Exercise 4.1
Run the code at **Step1-Euler** and try a variety of step lengths from 0.01 to 4.

4.2   DISCRETE TIME SOLUTION OF A FIRST-ORDER SYSTEM

There is a way to perform an accurate computer simulation with no limit on step size, provided that the input does not change between steps. To explain it, there is some fiddly mathematics that you might prefer to skip.

We will first consider the formal solution of the simple example. The treatment here may seem over elaborate, but later on we will apply the same methods to more demanding systems.

By using the variable x instead of H or T_{coffee} we can put all the simple examples into the same form:

$$\dot{x} = a\,x + b\,u \tag{4.1}$$

Here the dot over the x means *rate of change*, a and b are constants which describe the system while u is an input. This might simply be a constant in the same way that $T_{ambient}$ is in the coffee example.

Rearranging, we see that:

$$\dot{x} - a\,x = b\,u \tag{4.2}$$

Since we have a mixture of x and dx/dt, we cannot simply integrate the equation. We must somehow find a function of x and time which when differentiated will give us both terms on the left of the equation.

In the mathematical expressions that follow, we will use the stop '.' For multiplication, rather than the more cumbersome '*'.

Let us consider

$$\frac{d}{dt}\big(x.f(t)\big)$$

where $f(t)$ is some function of time which has derivative $f'(t)$. When we differentiate by parts we see that

$$\frac{d}{dt}\big(x.f(t)\big) = \dot{x}.f(t) + x.f'(t) \tag{4.3}$$

If we take Equation 4.2 and multiply it through by $f(t)$, we get

$$\dot{x}.f(t) - a.f(t).x = b.u.f(t) \tag{4.4}$$

Now if we can find $f(t)$ such that

$$f'(t) = -a.f(t)$$

then Equation 4.4 will become

$$\frac{d}{dt}\big(x.f(t)\big) = b.u.f(t)$$

and we will have something that we can integrate.

The solution to our mystery function is

$$f(t) = e^{-at}$$

and our equation becomes

$$\frac{d}{dt}\big(x.e^{-at}\big) = b.u.e^{-at} \tag{4.5}$$

When we integrate this between the limits 0 and t, we find

$$\left[x.e^{-at}\right]_0^t = \int_0^t b.u.e^{-at}\,dt$$

If u remains constant throughout the interval, we can take it outside the integral. We get

$$x(t).e^{-at} - x(0).1 = u\left[b.\frac{-1}{a}e^{-at}\right]_0^t$$

so

$$x(t).e^{-at} = x(0) + u\frac{b}{a}\left(1 - e^{-at}\right)$$

(4.6)

We can multiply through by e^{at} to get

$$x(t) = x(0).e^{at} + u\frac{b}{a}\left(e^{at} - 1\right)$$

(4.7)

This expresses the new value of x in terms of its initial value and the input in the form

$$x(t) = g\,x(0) + h\,u(0)$$

where

$$g = e^{at}$$

and

$$h = b\frac{e^{at} - 1}{a}$$

Now we are able to see how this relates to simulation with a long time-step. For steps of a constant width T, we can write $t = nT$. We then have a way of moving from one step to the next

$$x\big((n+1)T\big) = g\,x(nT) + h\,u(nT)$$

and the computer code to calculate the next value is simply

```
x = g*x + h*u
```

There is no "small dt" approximation, the expression is exact when the constants g and h have been calculated from the value of T.

The simulation program of the last section can now be made precise. In our calculation of g and h above, we replace a with the value $-k$ and b with value 1, so that for step length dt we have:

```
k = .5;      //leak coefficient
g = Math.exp(-k * dt); //Coefficients
h = (1 - g) / k;        // for an exact solution
```

```
x=0;                        //Initial level 0 to 40
u=15;                       //Input flow 0 to 20

t=0;
MoveTo(t,x);
while (t<tmax) {
    x = g*x + h*u;          //This is the simulation
    t = t + dt;
    Spot(t, x);             //This displays the result
}
```

The magic is all in that one line that calculates a new value for *x*. It almost looks too easy. But can the method cope with higher-order systems?

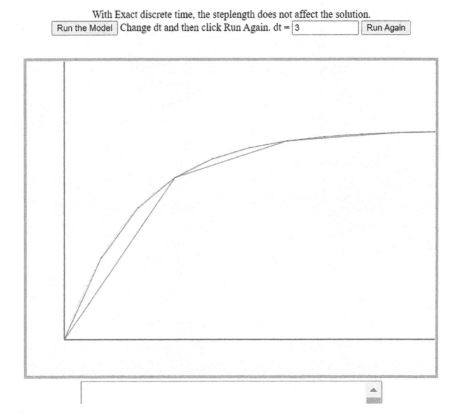

With Exact discrete time, the steplength does not affect the solution.
Run the Model | Change dt and then click Run Again. dt = 3 | Run Again

Screenshot of Step2-Exact.xhtml.

Exercise 4.2
Run the code at **Step2-Exact**. Use the 'Run again' button to try values of *dt* in the order 4, 2, 1, 0.5 and 0.1.

5 Modelling a Second-Order System

ABSTRACT

Now that we have got the groundwork prepared, we can start to use it to investigate a real problem. We start to see that a position controller is ruled by two 'state variables', its position and its velocity. The motor drive current can be made to depend on their values, in a 'feedback' process, that we hope will make the positioner do what we want it to.

5.1 A SERVOMOTER EXAMPLE

Now that we have got the basics of displaying the simulation results of the way, we can investigate the way that simulation relates to control theory.

A servomotor drives a robot axis to position x. The speed of the axis is v. The acceleration is proportional to the drive current u; at present there is no damping. Can we model the system to deduce its performance?

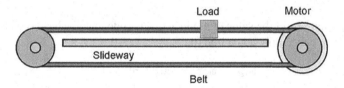

Position control example.

When we considered the water butt, we had an equation for the rate of change of the single variable that described the system. By repeatedly adding the small change over a short time interval, we were able to track the progress of the system against time. Let us try the same approach here.

One of the variables describing the system is its position x, so we must look for an equation for dx/dt. When we spot it, it is simply:

$$\frac{d}{dt}x = v$$

Clearly we have found a second 'state variable', velocity v, and we must now look for an expression for its rate of change, too. But we have been told that this acceleration is proportional to the input, u.

Our second equation is therefore

$$\frac{d}{dt}v = b.u$$

DOI: 10.1201/9781003363316-5

Now we have two differential equations, where their right-hand sides only contain variables that we 'know', so we can set up two lines of code

```
x = x + v*dt
v = v + b*u*dt
```

and we have only to 'top-and-tail' them and wrap them in a loop to make a simulation.

If we modify the code that we used to test step lengths, we have the core code

```
b=1;
dt=.01;
t=0;
x=-10;              //Initial position
v=0;
// set a value to u here

MoveTo (t, x);
while (t<10) {
  x = x + v*dt;   //This is the simulation
  v = v + b*u*dt;
  t = t + dt
  LineTo(t, x);
}
}
```

But we have not yet set the value of u. Since this is code that you have typed into your computer, you can simply add a line like

```
u=20;              //Input motor current
```

or

```
u = -4*x;
```

but to obey the security rules, we have to use text-box inputs to insert the feedback values.

You can see the code inside the source of **Motor1**. You will not be able to edit the code in the window, but three inputs have been added for you to vary the drive and the feedback.

As it first loads, the constant input rapidly drives the position off the screen as shown in Figure 5.1. The simulation only becomes interesting when we apply some position control.

Exercise 5.1
Make the drive proportional to the position error, by changing the numbers to make

```
u = -4*x;
```

When you click run, what do you see?

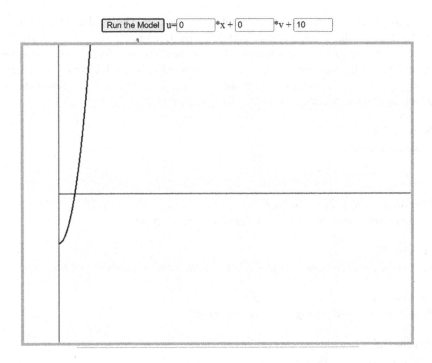

FIGURE 5.1   The plot before you apply feedback.

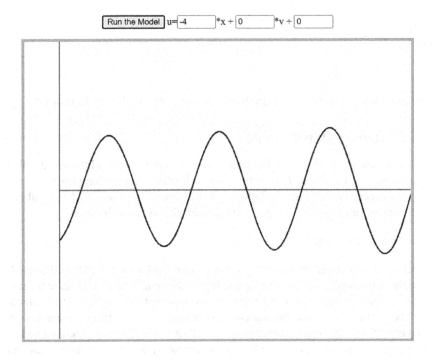

Motor1.xhtml with $u = -4x$.

Feeding back position alone just results in oscillation. The oscillations appear to grow because of the simulation error caused by the finite value of *dt*. You can see the effect of reducing its value. But for very small values there is a delay before the result appears. This is because the screen cannot be refreshed while the JavaScript is running.

To get stability we need to add some damping in the form of a velocity term. In fact we can try for a rapid response without overshoot.

Exercise 5.2

Change the input to:

```
u = -4*x - 2*v;
```

What does the response look like?

Then adjust the coefficient of *v* to get the 'best' response.

Exercise 5.3

What is the smallest value of that *v* coefficient that will avoid an overshoot?

Exercise 5.4

What is the result of setting the *v* coefficient to 20?

Exercise 5.5

What is the result if you set?

```
u = -64*x -14*v;
```

Exercise 5.6

Now try

```
u = -100*x -20*v;
```

It looks as though we can get an arbitrarily fast response with very little effort.

5.2   REAL-TIME SIMULATION

That earlier simulation showed the result almost instantly. With a little more embellishment, we can use the *setTimout* function to display a result in real time.

As the execution 'falls off the bottom' of the code in the window, the calculation ceases and any changes to the canvas are displayed. But the instruction

```
setTimeout ("loop();", 1000*dt);
```

has set an 'alarm clock' to wake up the code after a delay of 1000*dt* milliseconds. The code in the box is executed yet again, repeating for as long as *t* is less than *tmax*.

Now we can change inputs as the simulation runs, and even try to control it manually. The 'jollies' will now contain two text boxes, one for setting parameters and initial conditions, the other for executing at each time step. Instead of adding buttons for putting in step changes, with a little more ingenuity we can vary inputs continuously. The aim at first will be to keep things simple.

In the simulation **Motor2**, the windows have been split, although they are shown to display the code used, they cannot be edited. We will use this style of simulation for the investigation of drive limits that will follow.

The aim of the first setting of the variables is to convince you that control might not be easy!

Exercise 5.7

Run **Motor2** with the default values and try to drive its position by hand to zero. Drag the cursor along the input bar.

To make the task easier, the value of *dt* can be reduced.

Now we must investigate the effects of feedback. **Motor3** has been prepared with inputs for the position and velocity feedback, so that you can repeat the experiments that you tried on **Motor1**.

Exercise 5.8

Open **Motor3** and run it with the length *tmax* of the simulation set to ten seconds.

Set the feedback values to:

```
u = -4*x-2*v;
```

The result should look like Figure 5.2.

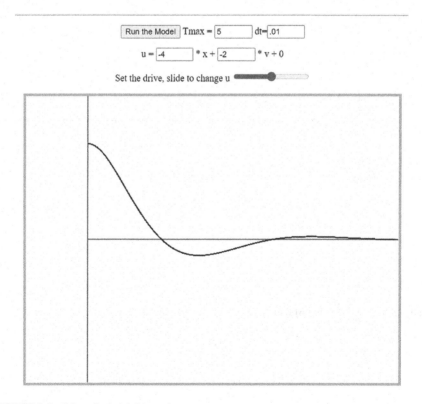

FIGURE 5.2   Motor2.xhtml, low gains.

Exercise 5.9

Now try

```
u = -64*x-10*v;
```

Exercise 5.10

Finally try (Figure 5.3)

```
u = -100*x - 20*v;
```

As we found before, it looks as though we can get an arbitrarily fast response with very little effort. We will soon be disillusioned!

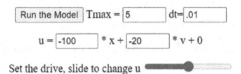

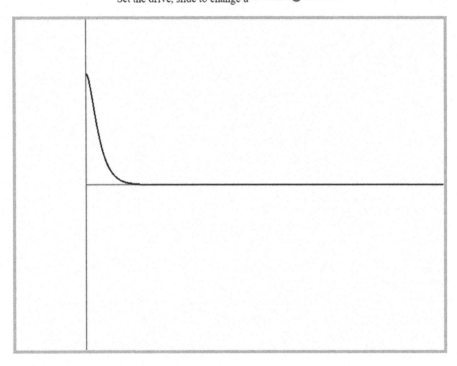

FIGURE 5.3   Motor2.xhtml with high gains.

6 The Complication of Motor Drive Limits

ABSTRACT

By the end of the last chapter, we were beginning to think that by using high feedback gains we could get the system to respond as quickly as we liked. This chapter brings us back to Earth by pointing out that the motor drive is limited. No longer are the rules of simple linear feedback enough.

The argument for high position gain may no longer be about settling time, it can be about resistance to a disturbing force. And to make that high gain give a good response, we must use velocity feedback far above the levels suggested by linear theory.

We see that the 'phase plane' is another way to visualise the performance, where velocity is plotted against position. We get a glimpse of 'sliding mode' in action.

6.1 DRIVE SATURATION

In the last section it looked as though we could obtain as fast a response as we like. But there is the problem that the motor drive is limited. Very few systems are totally linear, and we will see that for a good response the feedback must be nonlinear, too.

Consider our servomotor problem again. Now the input u is limited to lie between ± 1.0. When it is on either limit, the plot is red. The target position is set at zero.

Run the code at **Motor4** with an assortment of feedback gains, to compare the responses with those you found for **Motor3**.

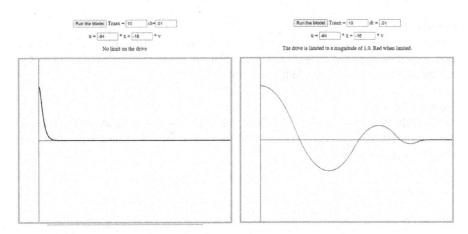

Comparison of Motor3.xhtml and Motor4.xhtml, showing the effect of drive saturation.

Exercise 6.1

Start with position gain = −64, velocity gain = −16. These values would give critical damping without the drive limitation.

You will see although these gains gave a very good response before, with the limited drive it is very different. The plot is coloured red wherever the drive has hit its limiting magnitude of 1. The response is dominated by those drive limits as much as by the other parameters of the system.

Exercise 6.2

Try position gain = −4, velocity gain = −6. Is that a little better with much lower gains?

Though the overshoot is much less, the system is slow to settle. But there is a much greater problem.

Exercise 6.3

Experiment with different values of feedback gains to try to get the fastest response that has no overshoot.

6.2 THE EFFECT OF A DISTURBANCE

Responses can look excellent on paper, but when applied to a practical system those gains might make it weak and ineffective. Not only must the controller bring the system to the desired position, it must be able to withstand disturbances.

The simulation at **Motor5** has been modified to let you apply a disturbing force. This will be added to the drive u to accelerate the system. The plot also shows the velocity in blue and the drive u.

The screenshot at Figure 6.1 shows that after settling, a disturbance of 0.4 of the maximum drive has been applied. With a position feedback gain of −4, this results in a position error of 0.1, meaning that position control will be 'soft'.

Exercise 6.4

Run it with gains for position −4 and velocity −2, and after the plot has settled on the axis, disturb it using the disturbance input. You can see that if this were an axis of a machine tool, the control would be very 'soft'.

Exercise 6.5

Now set the gains to position −100 and velocity −72. You will see that although the drive must change to counteract the disturbance, the effect on the position and velocity will be very much less.

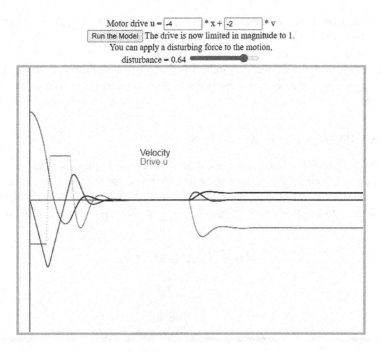

FIGURE 6.1 Screenshot of Motor5.xhtml with position gain 4, velocity gain 2.

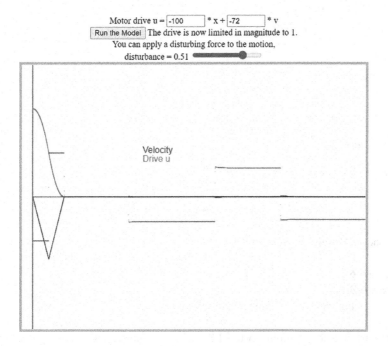

Screenshot of Motor5.xhtml, with position gain −100, velocity gain −72.

This is the reason that control engineers are eager to use a maximum loop gain where possible. By increasing the velocity coefficient to be much greater than the linear theory would recommend, crisp control is possible.

In a second-order continuous system, there are few limitations on the gain that can be applied, but the discrete-time considerations of computer control can cause instability.

6.3   A DIFFERENT VISUALISATION

The same sort of second-order position control task can be displayed with a simple 'move the blocks' simulation, written some time ago. The state equations are exactly the same, but instead of drawing a graph, an image is moved. Text boxes again allow the feedback gains to be set, but step changes of disturbance are now applied with buttons.

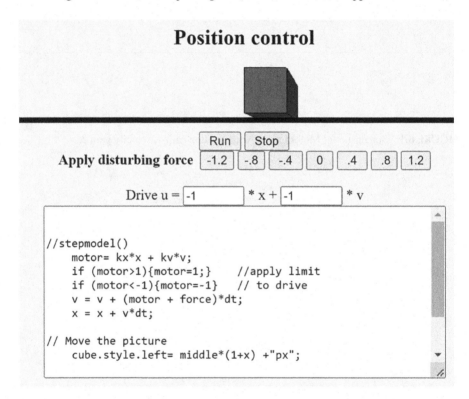

Screenshot of MoveBlock.xhtml.

Exercise 6.6
Run **MoveBlock**.

There is text beneath the image and text box:

You can adjust the damping kv to -1.8 to give a response that does not overshoot. But now apply a disturbance and see how 'soft' the control is.

Now try $kx = -100$ and $kv = -50$.

The block stops as though it has hit a rock.

Small disturbances have little effect – but of course a force of 1.2 N will overcome the motor and drive the block right off.

A conclusion is that for good resistance to disturbances, very large feedback gains must be used.

6.4   MEET THE PHASE PLANE

Another way to visualise the problem is to plot velocity against position. Such a plot is termed a phase plane.

Exercise 6.7

Phase1 to see it in action.

When you load it, it has default gains position -1, velocity -2. Later we will see that such values are examples of critical damping, where the position gain is ω^2 and the velocity gain is 2ω.

For these low values the limiter does not act.

Input u = [-1] * x + [-2] * v
Initial position = [-1]
The drive is now limited in magnitude to 1. The plot is red when u is on a limit.
[Run the Model]

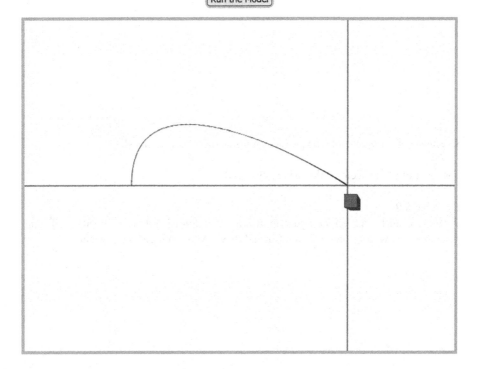

Screenshot of Phase1.xhtml with position gain -1 and velocity gain -2.

Exercise 6.8
Run again with pairs of values (−4, −4), then (−9, −6) and (−16, −8).

For this last pair of values, you will see that the drive limitation has caused an overshoot.

Input u = |-16 | * x + |-8 | * v
Initial position = |-1 |
The drive is now limited in magnitude to 1. The plot is red when u is on a limit.
[Run the Model]

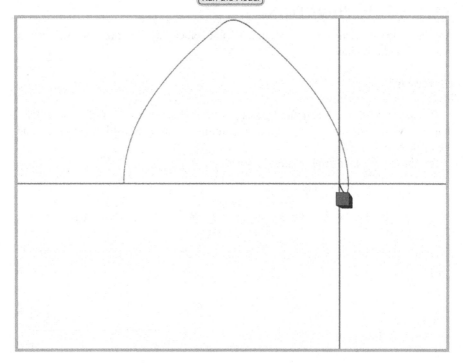

Screenshot of Phase1.xhtml with position gain −16 and velocity gain −8.

For most of the trajectory, the drive is limited.

Exercise 6.9
With the need for a high loop gain in mind, try position −100 and velocity −20. The response is now dominated by the limited drive. More velocity gain is needed.

Exercise 6.10

Now try the combination position −100, velocity −51.5.

You will see an example of time-optimal control. Full drive is used for both acceleration and deceleration, bringing the block to rest in minimum time.

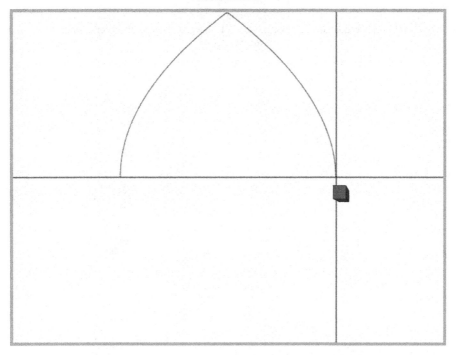

Input u = [-100] * x + [-51.5] * v
Initial position = [-1]
The drive is now limited in magnitude to 1. The plot is red when u is on a limit.
[Run the Model]

Screenshot of Phase1.xhtml with position gain −100 and velocity gain −51.5.

For that value of initial error, the control is time-optimal.

Exercise 6.11

For that same combination of gains, set the initial position to be $x = -0.5$.

You should again see the path start with full positive, then switch to full negative drive. But the trajectory then enters a straight-line section. This is called a *sliding mode*, in which the drive switches to and fro to keep the state on that line.

Here, $100*x + 51.5*v = 0$. This can be regarded as a differential equation in which the error decays to zero with a first-order time constant of just under two seconds.

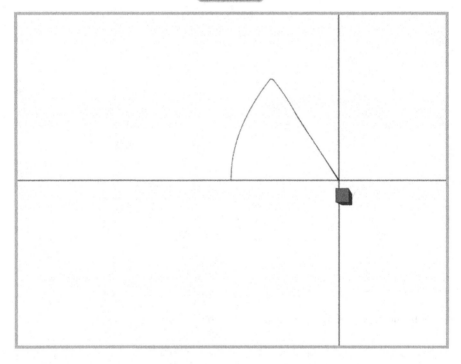

Input u = | -100 | * x + | -51.5 | * v

Initial position = | -0.5 |

The drive is now limited in magnitude to 1. The plot is red when u is on a limit.

[Run the Model]

Screenshot of Phase1.xhtml response to initial position of $x = -0.5$.

Exercise 6.12

Still keeping that combination of gains, change the initial conditions to $x = -1.5$;

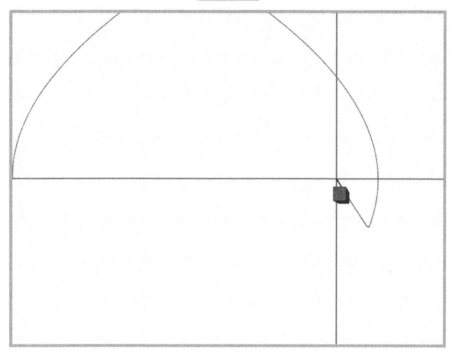

Input u = [-100] * x + [-51.5] * v
Initial position = [-1.5]
The drive is now limited in magnitude to 1. The plot is red when u is on a limit.
[Run the Model]

Screenshot of Phase1.xhtml, showing that an initial error of −1.5 gives an overshoot.

Once again you will see that there is an overshoot. That gain combination is only time-optimal for that particular magnitude of initial displacement. But we will soon see that there is an alternative way to consider feedback.

6.5   IN SUMMARY

From this sequence of exercises, you should have seen that the design of a practical control system must consider system constraints and nonlinearities as having the same importance as time constants. You might consider the following case study.

Exercise 6.13

You have been given the following design brief:

A servomechanism is in the form of a linear track. It must move a load of one kilogram a distance of one metre, coming to rest in one second. It must resist disturbances to the extent that a force of 10 N will only give a 1 mm deflection from its target position.

What are the steps in the design? Will a motor that can exert a force of 10 N be good enough?

7 Practical Controller Design

ABSTRACT

With drive limitation making our system nonlinear, gains that give a good response from one starting point can give an overshoot when starting from another. So we rearrange our calculation. A position error creates a demand for velocity to correct it. It is the difference between the demanded and actual velocity that determines the motor drive. By putting a limit on that velocity demand, the response is good from all starting points.

Can this simple tactic be employed to control higher-order systems? What is 'pragmatic control'?

If we assume that the speed is constant, a bicycle that is trying to follow a line has four state variables. I hope that the reader is experienced at riding a bicycle, since theory will be tied to intuition in analysing and simulating the task.

7.1 OVERVIEW

So far, we have only considered second-order systems. We have found that for 'stiff' control, high loop gains are needed to resist disturbances. Conventional thinking is to decide on a position gain, then to add appropriate velocity feedback to damp the response. But we find that when drive limitation comes into play, velocity gains must exceed any values suggested by a damping factor. Perhaps we should look at the velocity feedback gain first.

A first-order system, like speed control of a motor, is always stable however high the gain. So, the velocity feedback should be increased until other effects in the amplifier or simulation come into play. We find that when adding in the position term, there is an advantage in making it nonlinear. This leads to a successful strategy for higher-order systems, a technique which I choose to call '*Pragmatic Control*'.

7.2 THE *VELODYNE* LOOP

When we apply a large velocity feedback gain to a motor control system, we have a system that will respond briskly to a demanded change of speed, or strongly apply drive to combat any disturbing force. The result can be termed a 'velodyne loop'.

In the previous simulations, we tried a position gain of 100 with a velocity gain of just over 50. We calculated the drive as

```
u = -100*x - 50*v;
```

With a velodyne in mind, we could instead calculate u in two stages. First, we define a desired velocity for that value of position error, then we close the velocity loop.

```
vdem = -2*x;u = 50*(vdem - v);
```

DOI: 10.1201/9781003363316-7

Algebraically the result is exactly the same, but it gives us a different way to look at the problem. The position error is used to calculate a demanded speed, *vdem*. The loop then strongly applies drive to achieve that value of speed.

We have already seen that linear control can mean that the response is time-optimal for one value of initial error, but it will result in an overshoot for a larger one. We can improve performance by adding a nonlinearity to the feedback.

7.3   DEMAND LIMITATION

We saw that while gains of position −100 and velocity −51.5 gave the fastest possible response from an initial error of 1.0, an initial error of 1.5 caused an overshoot. How can we ensure that there is no overshoot for any size of error? The answer is to apply a limit to the demanded velocity *vdem*. Our controller has become nonlinear.

The feedback code for **Phase2** is now

```
vdem=-kx*x;
if (vdem>1){vdem=1;}
if (vdem<-1){vdem=-1;}
u=kv*(vdem - v);
```

The default values of the gains are $kx=2$ and $kv=51.5$.

kx now has a different meaning. It was previously the gain of the contribution to u. It is now the gain of the position's contribution to the demanded velocity *vdem*.

Exercise 7.1
Run **Phase2** and try various initial conditions. With the default values you again see time-optimal control, but when you put the initial value for $x=-1.5$ there is a severe overshoot.

If you now change the velocity limitation from 2 to 1, there will be no overshoot for values greater than unity.

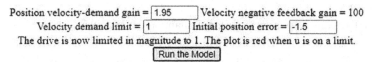

Position velocity-demand gain = 1.95   Velocity negative feedback gain = 100
Velocity demand limit = 1   Initial position error = -1.5
The drive is now limited in magnitude to 1. The plot is red when u is on a limit.

Run the Model

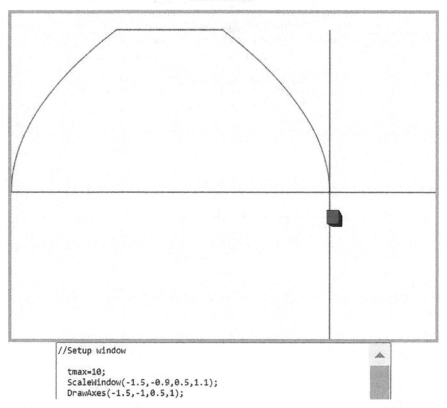

```
//Setup window

    tmax=10;
    ScaleWindow(-1.5,-0.9,0.5,1.1);
    DrawAxes(-1.5,-1,0.5,1);
```

Screenshot of Phase2.xhtml with speed demand limited to 1.

Exercise 7.2

What happens when you set the velocity limit to 0.5?

Now the response is much slower, ending with a sliding decay. But you can double the position gain to 4, and once again the final settling is optimal, unless the initial value was small.

Exercise 7.3

Now see the same responses in a time-domain plot at **Motor6**.

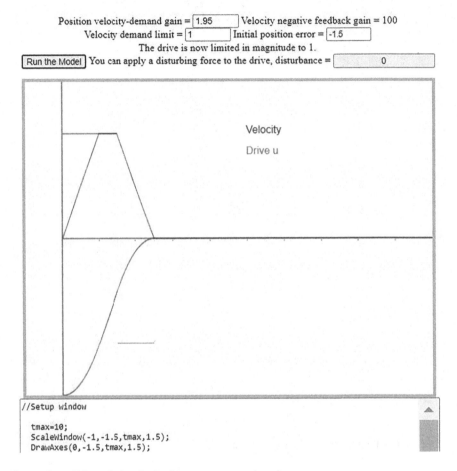

Position velocity-demand gain = 1.95   Velocity negative feedback gain = 100
Velocity demand limit = 1   Initial position error = -1.5
The drive is now limited in magnitude to 1.
Run the Model   You can apply a disturbing force to the drive, disturbance =   0

Velocity

Drive u

```
//Setup window

  tmax=10;
  ScaleWindow(-1,-1.5,tmax,1.5);
  DrawAxes(0,-1.5,tmax,1.5);
```

Screenshot of Motor6.xhtml, plotting responses against time.

7.4   RIDING A BICYCLE

Now here is an extremely difficult control problem that most of us have learned to master mentally. The position controller had two state variables, this has four. If you consider distance and speed, it has two more, but we will just consider the bicycle as travelling at constant speed along a track.

1. For our first variable, we will take x as the distance from the track centre line.
2. The second variable *ang* will be the heading angle of the bicycle, relative to that line.
3. Next is the angle at which we are leaning from vertical, *lean*.
4. Finally, we have the rate at which that tilt is changing, *leanrate*.

We will assume that the bicycle has length L between front and back wheels, and that it is travelling at constant speed V. We will also assume that angles are small.

So, our first task is to derive some equations for their rates of change.

The distance from the line will change according to the heading angle of the bicycle, so

$$d/dt(x) = V * \text{ang}$$

The heading angle will change at a rate proportional to the handlebar angle, *steer*.

$$d/dt(ang) = \text{steer} * V/L$$

It is easy to see that

$$d/dt(\text{lean}) = \text{leanrate}$$

but we will have to think hard to find the rate of change of leanrate, the acceleration of the lean angle.

The rider's body will accelerate horizontally according to the lean angle, $g*\sin(\text{lean})$.

The lean angle is the difference between the body position and the point where the tyre meets the ground, x, divided by the height H of the rider's centre of gravity.

Now x is accelerating at a rate proportional to steer,

$$\ddot{x} = \text{steer}\frac{V^2}{L}$$

so

$$d/dt(\text{leanrate}) = (g*\sin(\text{lean}) - \text{steer}*V^2/L)/H.$$

To tidy this up, let us give values to some of the symbols.

Let the wheelbase L be one metre, with a small rider one metre above the ground. We will approximate g to 10 and consider the bicycle to be travelling at 4 m/s, that is 14.4 km/hr.

So we have $L=1$, $H=1$, $g=10$ and $V=4$.

The equations become

$$d/dt(x) = 4*\text{ang}$$

$$d/dt(ang) = 4*\text{steer}$$

$$d/dt(\text{lean}) = \text{leanrate}$$

$$d/dt(\text{leanrate}) = 10*\sin(\text{lean}) - 16*\text{steer}$$

We are ready to set up our simulation.

Exercise 7.4

Run **Bike1**. When you press 'run' everything is perfectly balanced. Now drag the handlebar slider slightly. As soon as there is a slight movement, the bike will start to fall over. It is unlikely that you can ride it, even with practice.

Exercise 7.5

Now run **Bike2**. Time has been slowed down by a factor of 5 to give you a chance, but now the simulation will stop if you hit the ground. Although the tilt is very lively, the position will take a long time to react. For how long can you balance the bike?

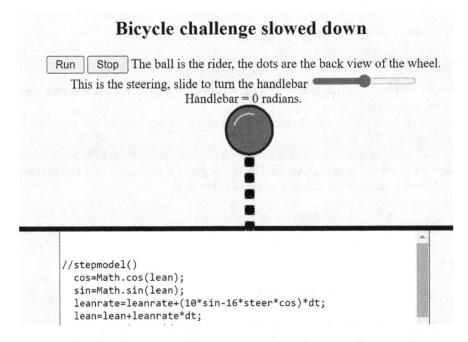

Bicycle challenge slowed down

Run   Stop   The ball is the rider, the dots are the back view of the wheel.
This is the steering, slide to turn the handlebar
Handlebar = 0 radians.

```
//stepmodel()
  cos=Math.cos(lean);
  sin=Math.sin(lean);
  leanrate=leanrate+(10*sin-16*steer*cos)*dt;
  lean=lean+leanrate*dt;
```

Screenshot of Bike2.xhtml.

What do you learn about riding the bike? The first requirement is not to fall off. If you are falling to the right, you turn the handlebar swiftly to the right, but it is easy to overdo the correction. In any feedback it is necessary to include *leanrate* as well as *lean* to add some damping.

But it is also worth noting that the feedback coefficients are positive. In a position controller they are negative.

So to try to keep upright, we will have a line of code such as

```
steer = leangain*lean + leanrategain*leanrate;
```

How should we choose values for those gains?

The simulation uses sine and cosine to give realism to the image, but if the control works, the lean angle should remain small. When linearised, the code line

```
leanrate=leanrate+(10*sin-16*steer*cos)*dt;
```

becomes

```
leanrate=leanrate+(10*lean-16*steer)*dt;
```

It tells us that the lean gain must be greater than 10/16 for the steering to overcome the tendency to fall over. You will know from your own experience that if cycling slowly, you have to turn the handlebars more vigorously. So we might try a value of 20, adding in 2 × leanrate to damp the response.

Now that our handlebars are to be turned automatically, we can use the slider bar to set a demanded lean angle, *leandem*.

```
steer=20*(lean-leandem)+2*leanrate;
```

That's all that it takes to tame the system.

Exercise 7.6
Run **Bike3**. This now automatically turns the handlebar to stay upright, and your slider is a demand for lean angle. The calculation has been restored to match real time. You will now see that when you lean to the left, your track turns to the left, just as you experience on a real bike.

The next step towards full line-following control is to control the heading angle. From our heading error we calculate a demanded lean angle. To turn to the left, you lean to the left. But if the lean is too big, you will fall over. So we might have

```
leandem= 1.0*(angdem-ang);
if(leandem>0.3){leandem=0.3;}
if(leandem<-0.3){leandem=-0.3;}
```

But where did *angdem* come from? If we are off the line, we want to steer towards it, so we demand an angle proportional to the error. But if we are a long way off the line we might then ride in circles, so again we need to limit the demand to half a radian, about thirty degrees.

```
angdem=0.4*(xtarget-x);
if(angdem>0.5){angdem=0.5;}
if(angdem<-0.5){angdem=-0.5;}
```

And if you do not want the handlebar to band into your knee, you want to limit that to half a radian, too. But to simplify the code, it is worth defining a function limit(*a*, *b*), so that we can write

```
angdem=limit(0.4*(xtarget-x), 0.5);
leandem=limit(1.0*(angdem-ang), 0.3);
steer=limit(20*(lean-leandem)+6*leanrate, 0.5);
```

Exercise 7.7

These have been built into **Bike4**, where performance is improved by using rather more *leanrate* feedback than before. But I am sure that the performance can be greatly improved by tuning the rather arbitrarily chosen parameters. Run it and test the result.

A view of all the variables changing in time will give a clearer view of what is happening.

Exercise 7.8

Run **Bike5** to see a plot of the variables. x is plotted in black heading angle in blue and lean in red.

7.5   NESTED LOOPS AND PRAGMATIC CONTROL

Do you see a pattern building here? In a high order system, loops are closed one after another, with demands linking one loop to the next. By limiting the magnitude of these demands, we have a nonlinear controller that gives good performance over a large range.

Control theory is something of a fashion industry. At one time great effort was put into developing variable structure control, then came fuzzy control, where the 'fuzzifier' is merely a way of generating a piecewise linear nonlinearity. Genetic algorithms were once all the rage. Neural nets are still in fashion, a way to tune a nonlinear function of numerous inputs.

But the essential message is the realisation that although analytic methods go a long way towards giving insight for linear controllers, logically devised nonlinearities can improve their performance.

8 Adding Dynamics to the Controller

ABSTRACT

So far, we have not asked our controller to be very clever, just adding the sensor signals together and maybe applying limits. But what if an essential sensor signal is missing, such as velocity? Can we add dynamics to our controller to estimate that signal and restore stability?

We can simulate a method that might be used by an analogue controller, or we can realise that our computer works in discrete time. We can simply estimate velocity from the change of position since the previous iteration.

8.1 OVERVIEW

As I have already said, there are two types of controllers. Simple ones feed back a mixture of sensor signals as soon as they are measured. Others add their own dynamics to the system, often to estimate a missing velocity or rate of change.

In theory, a first-order system, such as speed control of a motor, is always stable, however high the gain. If the control is discrete time, however, there is a limit to that gain. All will be revealed when discussing poles and zeros.

Exercise 8.1

Revisit **Motor3**. Keep the velocity gain at zero and view the response by setting the position gain in turn to 0.25, then 1.0 and 4.0. You will see that at each change the frequency doubles. The oscillation grows slightly because of the Euler approximation.

DOI: 10.1201/9781003363316-8

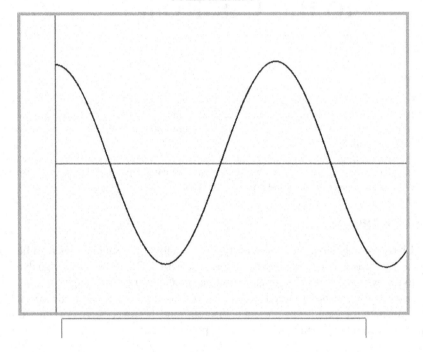

Screenshot of Motor3.xhtml with zero damping.

If we only have a position sensor, how can we apply damping? We must *estimate* the velocity.

The most obvious way to do this is to take the change in *x* over the last interval and divide it by *dt*. So, we must remember the previous position as *xold*.

Before updating *x*, we remember it with

```
xold=x;
```

and after updating, we calculate the estimated velocity *vest* with

```
vest=(x-xold)/dt;
```

If the 'eval' function had been permitted, you could have edited the code in the textarea to try it. But that is not really necessary. If you sort out the algebra, the differencing just pulls out the same value for *vest* as the actual value of *v*. The simulation would work exactly as before.

That is fine for a simulation, but it runs into trouble on a real physical system.

8.2   NOISE AND QUANTISATION

In a real system, it is necessary to use some sort of transducer to measure the position. In a digital system, this will be quantised, where finer resolution will cost more. We might, however, use an *incremental encoder*, where stripes moving across a photocell or magnetic sensor are counted.

We can simulate a transducer with a resolution of 0.01 by using

```
xq=Math.round(100*x)/100;
```

to produce the quantised signal *xq*.

Exercise 8.2
See how that performs in **Motor7**.

The velocity signal is now quantised too. When the velocity is small it just consists of occasional ones and zeros. But the controlled system is now just as stable as we would wish.

If we need a smoother velocity signal to display, there is another way to calculate it. We follow the example of the linear circuit, where a capacitor generates a 'lagged' version of the position, while the velocity is calculated from the difference between *x* and its lagged version.

```
xest=10*(x-xslow);
xslow=xslow+10*(x-xslow)*dt;
```

Exercise 8.3
Run **Motor8** to see the difference.

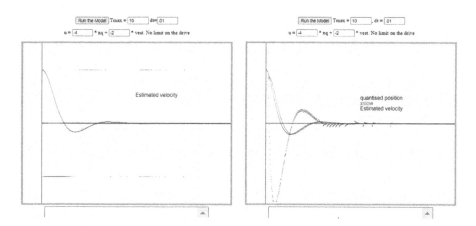

A comparison of velocity estimates between Motor7.xhtml and Motor8.xhtml.

8.3   DISCRETE TIME CONTROL

So far, we have been assuming that control can be applied as fast as our simulation runs, a hundred times per second. But assume now that we can only read the states and change the input five times per second. How do we build that into the simulation?

We can set a variable *tnext* to be the time to apply the control. Updating the input only happens if *t>tnext*. So, if we assume that the velocity signal is available, we will have code:

```
If(t>tnext){
   tnext=tnext+interval;
   u=kx*x+kv*v;
}
```

Of course, we must set an interval value if we want five updates per second.

Exercise 8.4

Run **Discrete1** and try combinations of feedback gains. You can also change the interval.

When the gains are low, discrete time makes little difference. But as we try successive pairs which would normally give deadbeat control, −4, then −16, −8 and −25, −10, the problems show up. Nevertheless, a combination of gains such as 20, 7 can give a remarkably good response when the interval is 0.2 second.

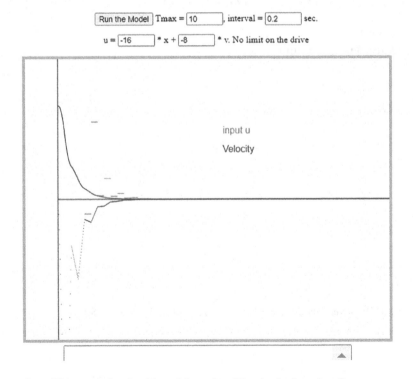

Screenshot of Discrete1.xhtml, with position gain −16 and velocity gain −8.

But suppose now that we have no velocity signal. We must estimate it from the position signal that is only available five times per second. We will now be very limited in the feed-back that we can use.

In the method that we used earlier, we estimated the velocity by subtracting from *x* its previous value, then dividing by *dt*. Now that previous value was taken at the last sample, so we have to divide the difference by *interval*.

Exercise 8.5
Run **Discrete2** and adjust the feedback values for the best performance you can get. Can you find a reasonable value for velocity gain if the position feedback factor is greater than four?

In a later section we will see how to apply some mathematical analysis to this.

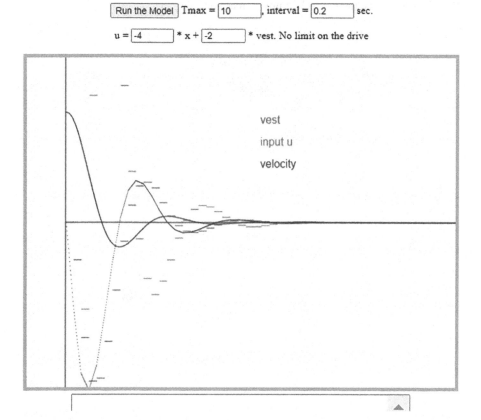

Screenshot of Discrete2.xhtml, showing the effect of discrete estimated velocity.

8.4   POSITION CONTROL WITH A REAL MOTOR

So far, we have considered position control to be ruled by two simple integrations. But in a real DC motor, the acceleration drops off with velocity until the motor reaches a steady speed. We have

$dv/dt = u - d*v$

where d is the natural damping. The motor will have a time constant $1/d$ and for constant input will reach a maximum speed of u/d.

With a limit on the drive, this means that a velocity feedback value can be found to give optimal stopping from any initial position.

Exercise 8.6
Run **Phase3** to see the natural damping in action.

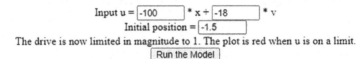

The drive is now limited in magnitude to 1. The plot is red when u is on a limit.

Screenshot of Phase3.xhtml, showing the effect of motor natural damping.

8.5 IN CONCLUSION

This should have given you an insight into the methods and problems of controlling a real system, although it has been centred on simulation. The treatment has been empirical, avoiding theoretical or mathematical treatments. Those are to come! Examples were chosen here where a nonlinear approach would have greatest advantage. However, there are numerous types of system, such as the stabilising of a chemical process, where advanced linear analysis can win the day.

9 Sensors and Actuators

ABSTRACT

Now we take a rest from the computing aspects and consider the hardware of a control system. Variables do not come out of thin air, but must be measured by a sensor of one sort or another. Motors require the means to apply their computed drive current. The selection is biased towards the sort of mechatronic system that could be a simple laboratory experiment.

9.1 INTRODUCTION

The previous chapters have been largely about simulation and feedback, examining the consequences of applying output signals to the input of a system. But if the system is real, how do we obtain these outputs and how do we apply them?

An electric iron manages to achieve temperature control with one single bimetal switch. To guide an autonomous car requires rather more control complexity. Control systems can be vast or small, they can aim at smooth stability or can achieve their objective with a switching limit cycle. They can be designed for supreme performance or can be the cheapest and most expedient way to control a throwaway consumer product. So where should the design of a controller begin?

There must first be some specification of the performance required of the controlled system. In the now-familiar servomotor example, we must be told how accurately the output position must be held, what forces might disturb it, how fast and with what acceleration is the output position is required to move. Considerations of reliability and lifetime must then be taken into account. Will the system be required to work only in isolation, or is it part of a more complex whole?

A servomotor for a simple radio-controlled model will use a small DC motor, with a potentiometer to measure output position. For small deviations from the target position, the amplifier in the loop will apply a voltage to the motor proportional to error, and with luck the output will achieve the desired position without too much overshoot.

An industrial robot arm requires much more attention. The motor may still be DC, but it will probably be of high performance at no small cost. To ensure a well-damped response, the motor may well have a built-in tachometer that gives a measure of its speed. A potentiometer is hardly good enough, in terms of accuracy or lifespan. An incremental optical transducer is much more likely – although some systems have both. Now, it is unlikely that the control loop will be closed merely by a simple amplifier; a computer is almost certain to get into the act. Once this level of complexity is reached, position control examples show many common features.

When it comes to the computer that applies the control, it is the control strategy that counts, rather than the size of the system. A radio-telescope in the South of England used to be controlled by two mainframes, but had dubious success. They were replaced by two cheap personal microcomputers, and the performance was much improved.

DOI: 10.1201/9781003363316-9

9.2 THE NATURE OF SENSORS

Without delving into the physics or electronics, we can see that there are character-istics of sensors and actuators that are fundamental to many of the control decisions. Let us try to put them into some sort of order of complexity.

By a factor of orders of magnitude, the most common sensor is the thermostat. Not only is it the sensor that detects the status of the temperature, it is also the controller that connects or disconnects power to a heating element. In an electric kettle, the operation is a once-and-for-all disconnection when the kettle boils. In a hot-water urn, the connection re-closes to maintain the desired temperature as it cycles between limits.

Even something as simple as this needs some thought. To avoid early burnout of the contacts, the speed of the on–off switching cycle must be relatively slow, some-thing usually implemented by hysteresis in the sensor. Electric cooking rings are controlled by an *energy regulator* that applies switched on–off control, and there are regulations limiting how fast it is allowed to switch.

Many other sensors are also of a 'single bit' nature. Limit switches, whether in the form of microswitches or contactless devices, can either bring a movement to an end or can inhibit movement past a virtual end stop. Similar sensors are at the heart of power-assisted steering. When the wheel is turned to one side of a small 'backlash' zone in the mechanical link from wheel to steering, the 'assistance' drives the steer-ing to follow the demand, turning itself off when the steering matches the wheel's demand.

The classical way to signal the value of a continuous measurement is by means of a varying voltage or current. A common position transducer is the potentiometer. A voltage is applied across a resistive track and a moving 'wiper' picks off a voltage that is proportional to the movement along the track. This is shown in Figure 9.1. There is no quantisation of the voltage, but in any but the most expensive of poten-tiometers there will be 'gritty' noise as the wiper moves. There is also likely to be nonlinearity in the relationship between voltage and position.

Non-contact variations on the potentiometer principle include linear Hall effect devices that sense the varying angle of a magnetic field. There are also

FIGURE 9.1   A potentiometer can measure position.

transformer-based devices such as the "E and I pickoff" or the "LVDT" ('Linear Variable Differential Transformer') in which movement changes the coupling between an alternating field and a detection coil.

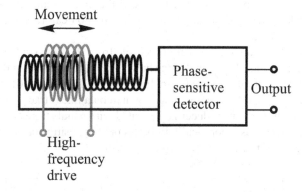

LVDT displacement sensor.

A method of measuring an angle is based on the Hall effect. A semiconductor device returns a voltage that varies with the strength of a magnetic field perpendicular to its surface. With two such sensors at right angles, the sine and cosine of the angle of an actuating magnet can be measured, as shown in Figure 9.2.

There are many Hall effect devices on the market, some giving a linear output, others giving a switched output as a contactless substitute for a microswitch limit switch.

To measure force, a strain gauge relies on the variation of the resistance of a thin metallic film as it is stretched. It is just one of the many devices in which the quantity is measured by means of a displacement of one sort or another, in this case by stretch.

Even though these measurements might seem to be continuous and without any steps, quantisation will still be present if a digital controller is involved. The signal must be converted into a numerical value. The analogue-to-digital converter might be '8 bit', meaning that there are 256 steps in the value, or up to 16 bits with 65,536 different levels. Although more bits are possible beyond this, noise is likely to limit their value.

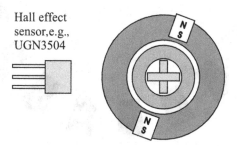

FIGURE 9.2   Hall effect transducer measuring sine and cosine.

9.3 THE MEASUREMENT OF POSITION AND DISPLACEMENT

An alternative way to measure a displacement is an *'incremental encoder'*. The signals are again based on simple on–off values, but the transitions are now counted in the controller to give a broad range of results. If the motion can only be in a single direction, simple counting is adequate. But if the motion can reverse, then a 'two-phase' sensor is needed.

A second sensor is mounted quarter of a cycle from the first, so that the output is obtained as pairs of bits. The signals are shown in Figure 9.3. Now the sequence 00 01 11 10 will represent movement in one direction, while 00 10 11 01 will represent the opposite direction. The stripes can be mounted on a linear track, just as easily as on a rotating motor. If the transducer is presented with equal stripes at one-millimetre intervals, the length is theoretically limitless but the position cannot be known to better than within half of each half millimetre stripe.

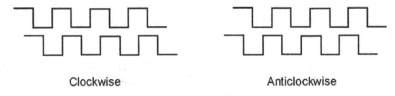

Clockwise   Anticlockwise

2-phase encoder waveforms.

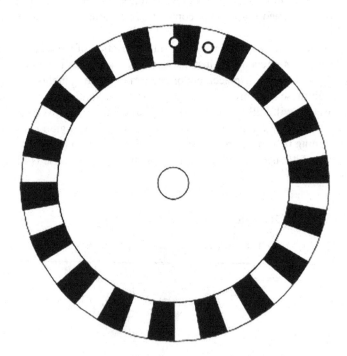

FIGURE 9.3   Stripes with two sensors for an incremental encoder.

This type of sensor has the advantage of extreme simplicity and was at the heart of the old 'rolling ball' computer mouse. Now it is built into many commercial DC motors. However, it has the disadvantage that the sensing is of relative motion, and there is no knowledge of the position at switch on. It is a displacement measurement, not a position measurement. To translate the reading into a position, it must be calibrated by driving it to some known position, either an end stop or a sensor.

A more subtle technique was used in the Puma robot. As well as the incremental stripes, of which there were around a hundred per revolution of the geared motor, an extra stripe was added so that one specific point in the revolution could be identified. At start-up, the motor rotated slowly until this stripe was detected. This still left tens of possibilities for the arm position, as the motor was geared to rotate many times to move the arm from one extreme to the other. But in addition to the incremental transducer, the arm carried a potentiometer. Although the potentiometer might not have the accuracy required for providing the precision demanded of the robot, it was certainly good enough to discriminate between whole revolutions.

The concept of 'coarse fine' measurement is a philosophy in itself!

For an instant measure of position, many stripes can be sensed in parallel. These are arranged as a 'Gray code', where only one stripe changes status at a time.

Figure 9.4 shows a disk giving five binary digits, that is 32 values per revolution. By increasing the number of sensors to ten, the rotation angle of a disk can be measured to one part in a thousand. But the alignment of the stripes and sensors must be of better accuracy than one part in a thousand. This results in a very expensive transducer.

FIGURE 9.4 Rotary Gray code.

9.4   VELOCITY AND ACCELERATION

As we have seen, it is possible to measure or 'guess' the velocity from a position measurement. The crude way is to take the difference between two positions and divide by the time between the measurements. A more sophisticated way is by means of the high-pass filter that we will analyse later.

Other transducers can give a more direct measurement. When a motor spins, it generates a 'back emf', a voltage that is proportional to the rotational velocity. Thus, the addition of a small motor to the drive motor shaft can give a direct measurement of speed. This sensor is commonly known as a *tachometer* or *'tacho'*.

The rotation of a moving body such as an aircraft is measured by a *rate gyro*. This is used to take the form of a spinning rotor that acted as a gyroscope. A rotation, even a slow one, would cause the gyroscope to precess and twist about a perpendicular axis. This twist was measured by an "E and I pickoff" variable transformer. Today, however, the 'gyro' name will just be a matter of tradition. In a tiny vibrating tuning fork, rotation about the axis causes the tines to exhibit sideways vibrations. These can be picked up to generate an output.

So 'chips' have become part of the hobby market, containing sensors for angular velocity, magnetic field angle and 3D gravitational acceleration. Even devices such as cameras and GPS transducers, time-of-flight and ultrasonic distance transducers should be considered as bought-in components that can complete a controller design.

9.5   OUTPUT DEVICES

When the output is to be a movement, there is an almost unbelievably large choice of motors and ways to drive them.

In the control laboratory, the most likely motor that we will find is the permanent magnet DC motor. These come in all sizes and are widely used in their automotive applications, such as windscreen wipers, window winders and rear-view mirror adjusters. Although a variable voltage could be applied to drive such a motor at a variable speed, the controller is more likely to apply a 'mark-space' drive. Instead of switching on continuously, the drive switches rapidly on and off so that power is applied for a proportion of the time.

To avoid the need for both positive and negative power supplies, a circuit can be used that is called an 'H-bridge'. The principle is shown in Figure 9.5. With the motor located in the cross-bar of the H, either side can be switched to the single positive supply or to ground, so that the drive can be reversed or turned off. If A and D on, while B and C off, the motor will be driven one way, while B and C on with A and D off will drive it the other way. By switching B and D on while A and C are off, braking can be applied to stop the motor. The bridge can be constructed from discrete MOSFET transistors, though bridges are also available as complete semiconductor components.

Another hobbyist favourite is the 'stepper motor'. There are four windings that can be thought of as North, South, East and West. Each has one end connected to the supply, so that it can be energised by pulling the other pin to ground. Just like a compass, when the North winding is energised, the rotor will move to a North position.

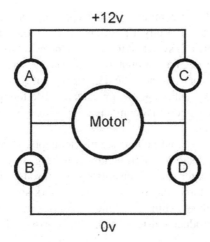

FIGURE 9.5   Schematic of H-bridge.

The sequence N, NE, E, SE, S, SW, W, NW and back to North will move the rotor through a cycle to the next North position. That might only move the motor through a fraction of a physical rotation, because there can be many poles in stepper motors. In a typical motor, it will take fifty of these cycles to move the rotor through one revolution.

Stepper motors are simple in concept, but they are lacking in top speed, acceleration and efficiency. A simple demonstration at **Stepper** shows why. They were useful for changing tracks on floppy disks, when these were still floppy. Large and expensive versions are still used in some machine tools, but a smaller servomotor can easily outperform them.

Exercise 9.1

Run **Stepper** to see the problems as speed becomes large.

In order to drive a stepper motor in a laboratory experiment, it used to be possible to output a pattern of bits to a printer port, but today's computers do not have printer ports anymore. Instead, it is possible to connect a microcontroller to a USB port and send the pattern as a serial byte to a virtual COM port. Computers do not have real serial ports anymore, either.

So after 'topping and tailing', the Arduino code to perform the task would be something like:

```
void loop() {
  if (Serial.available()>0) {
  // if command byte has arrived
      command=(int)Serial.read();
      PORTC=command & 15;
  }
}
```

Of course, the output bits of the port must be buffered to give enough drive current.

A DC motor has a lot in common with a stepper motor, in that a magnetic field is acting to pull the rotor to a target position. As the motor approaches this stable position, however, the field is switched to pull it to the next target, urging it forwards. Simple motors use brushes for commutation, delivering current to conductors on a wound rotor that rotates in a magnetic field imposed by the stator. 'Brushless' motors use electronics to do the switching, to gain extended life for applications that are as simple as cooling fans for personal computers. In those devices, it is the permanent magnet rotors that rotate.

Electric motors are far from being the only actuators. Hydraulic and pneumatic systems allow large forces to be controlled by switching simple solenoid valves.

In general, the computer can exercise its control through the switching of a few output bits, where finer forces can be controlled by mark-space switching of a binary output. With a very high switching frequency, the bytes in a.wav file even suffice for driving speakers to reproduce music, though some purists will disagree.

10 Analogue Simulation

ABSTRACT

While taking a rest from digital matters, we can look at the task performed by the 'operational amplifier' in analogue controllers. Though overwhelmed by an avalanche of digital devices, there are still things that an analogue circuit can do that a digital one cannot.

By considering the simulation of a second-order system, our mind is led to state equations. We take an adventurous step towards exploiting matrices to manipulate and analyse them.

10.1 HISTORY

It is ironic that analogue simulation "went out of fashion" just as the solid-state operational amplifier was perfected. Previously the integrators had involved a variety of mechanical, hydraulic and pneumatic contraptions, followed by an assortment of electronics based on magnetic amplifiers or thermionic valves. Valve amplifiers were common even in the late 1960s and required elaborate stabilisation to overcome their drift. Power consumption was high, and air-conditioning essential.

Adding to the cost of analogue computing was that of achieving tolerances of 0.01% for resistors and capacitors, and of constructing and maintaining the large precision patch boards on which each problem was set up.

Before long an operational amplifier became available on a single chip, then four to a chip, at a price of a few cents. It was now easy to build dynamics into a controller, just by adding such a chip with a few resistors and capacitors. But for simulation it must now be deemed easier and more accurate to simulate a system on a digital computer.

10.2 ANALOGUE CIRCUITRY

An area where the analogue circuit holds its own is in 'signal conditioning'. That does not just mean amplifying the feeble voltage from a strain gauge, it can be used for 'filtering'. When the signals to be digitised have sinusoidal noise on them at a frequency near the sampling rate, there is a phenomenon called 'aliasing'. The digitised signal seems to carry a sinewave at a much lower frequency. If you can remember seeing the wagon wheels in old westerns, turning slowly backwards while the wagon is hurtling at speed, you have seen aliasing. No amount of digital filtering will remove it, but with the aid of a resistor and a capacitor, maybe backed up by an operational amplifier, the high frequency is removed before it is digitised.

In the laboratory, the analogue computer still has its uses. Leave out the patch board and solder up a simple problem directly. Forget the 0.1% components – the parameters of the system being modelled are probably not known closer than a percent or two, anyway. Add a potentiometer or two to set up feedback gains, and a lot of valuable experience can be acquired. Take the problem of position control, for example.

DOI: 10.1201/9781003363316-10

An analogue integrator is made from an operational amplifier by connecting the non-inverting input to a common rail, analogue ground, while the amplifier supplies are at +5 volts and −5 volts relative to that rail (more or less). The inverting input is now regarded as a *summing junction*. A feedback capacitor of, say, 10 microfarads feeds back the output to this junction, while inputs are each connected via their own 100 kilohm resistor. Such an integrator will have a *time constant* of one second. If +1 volt is applied at one input, the output will change by one volt (negatively) in one second.

To produce an output that will change in the same sense as that of the input, we must follow this integrator with an invertor, a circuit that will give −1 volt output for +1 volt input. Another operational amplifier is used, also with the non-inverting input grounded. The feedback to the summing junction now takes the form of a 100 kilohm resistor, and the single input is connected through another 100 kilohm resistor as shown in Figure 10.1.

The gain of a virtual-earth amplifier is deduced by assuming that the feedback always succeeds in maintaining the summing junction very close to the analogue ground, and by assuming that no input current is taken by the chip itself. The feedback current of an integrator is then $C\,dv/dt$, where v is the output voltage. The input current is v_{in}/R, and so

$$C\frac{dV}{dt} + \frac{1}{R}v_{in} = 0$$

Now if the integrator's 10 microfarad feedback is reduced to 1 microfarad, its time constant will only be 0.1 second. If instead the resistor R is doubled to 200 kilohms the time constant will also be doubled. Various gains can be achieved by selecting appropriate input resistors.

Now in our position control example we have two "state variables" x and v, governed by:

$$\frac{dx}{dt} = v$$

and

$$\frac{dv}{dt} = bu$$

If we set up two integrators, designating the outputs as x and v, and if we connect the output of the v integrator to the input of the x integrator, and the input signal u to

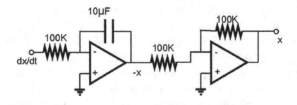

FIGURE 10.1 Construction of a non-inverting integrator.

the input of the v integrator through an appropriate resistor, then the simulation of Figure 10.2 is achieved.

As in the case of the digital simulation, the exercise only becomes interesting when we add some feedback.

As in the digital simulation, we want to make

$$u = f.(x_{\text{demand}} - x) - d.v$$

where x_{demand} is a target position, f is the feedback gain and d applies some damping.

If we connect a potentiometer resistor between the outputs $+v$ and $-v$ of the operational amplifiers relating to v, then the wiper can pick off a term of either sign as shown in Figure 10.3. Thus, we can add positive or negative damping. Another potentiometer connected between the supplies can represent the position demanded, while a further one feeds back the position error. The potentiometer signals are mixed in an invertor (which doubles as a summer), giving an output u which is applied to the input of the v integrator.

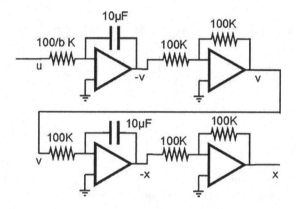

FIGURE 10.2 Simulating a second-order system.

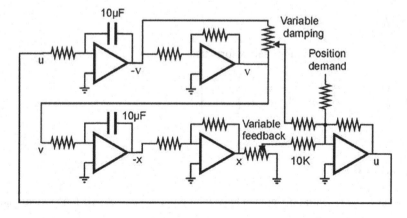

FIGURE 10.3 Second-order system with variable feedback.

10.3   STATE EQUATIONS

Although we have met state variables in an informal way, mathematics has been avoided. But now it is necessary to formalise them with *matrix state equations*.

In both the digital and analogue cases, we saw that the differential state equations merely list the inputs to each integrator in terms of system inputs and the state variables themselves. If we look at the integrator inputs of our closed-loop system, we see that dx/dt is still v, but now dv/dt has a mixture of inputs.

When we substitute the feedback value $f.(x_{\text{demand}} - x) - d.v$ for u, we find that the equations have become:

$$\frac{dx}{dt} = v$$

and

$$\frac{dv}{dt} = b\left(f.(x_{\text{demand}} - x) - d.v\right)$$

These equations can be expressed as a single matrix equation

$$\begin{bmatrix} \dot{x} \\ \dot{v} \end{bmatrix} = \begin{bmatrix} 0 & 1 \\ -b.f & -b.d \end{bmatrix} \begin{bmatrix} x \\ v \end{bmatrix} + \begin{bmatrix} 0 \\ b.f \end{bmatrix} x_{\text{demand}}$$

which has the form

$$\dot{\mathbf{x}} = \mathbf{A}\mathbf{x} + \mathbf{B}\mathbf{u}$$

where in this case the input $\mathbf{u}$ is x_{demand}.

Now compare this with our original equations before we considered feedback,

$$dx/dt = v$$

$$dv/dt = b.u$$

These can also be written in matrix form as

$$\begin{bmatrix} \dot{x} \\ \dot{v} \end{bmatrix} = \begin{bmatrix} 0 & 1 \\ 0 & 0 \end{bmatrix} \begin{bmatrix} x \\ v \end{bmatrix} + \begin{bmatrix} 0 \\ b \end{bmatrix} u$$

with the same form as

$$\dot{\mathbf{x}} = \mathbf{A}\mathbf{x} + \mathbf{B}\mathbf{u}$$

So, what is the essential difference between open loop and closed-loop equations?

11 Matrix State Equations

ABSTRACT

Now we must roll up our sleeves and get serious about the way that matrix algebra can be brought to bear on a control problem. These formal methods are seen to be equivalent to ad hoc techniques when the problem is simple, but they can come into their own when computing is brought to bear.

11.1 INTRODUCTION

We have seen that states, such as present position and velocity, can be used to describe the behaviour of a system, and that these can be woven into a set of first-order differential equations. It is tempting to assert that every dynamic system can be represented by a set of state equations in the form:

$$\dot{x} = Ax + Bu$$

where **x** and **u** are vectors and where there are as many separate rows as **x** has components.

Unfortunately, there are many exceptions. A sharp switching action cannot reasonably be expressed by differential equations of any sort. A highly nonlinear system will only approximate to the above form of equations when its disturbance is very small. A pure time delay, such as that of water travelling through the hose of your bathroom shower, will need a variable for the temperature of each drop of water that is in transit. Nevertheless, the majority of those systems with which the control engineer is concerned will fall closely enough to the matrix form that it becomes a very useful tool indeed.

An important property of this representation is that the matrix **A** can be analysed to find much about the system behaviour. It tells of stability and the time with which the system will recover from a disturbance. Presumably the system that we start with does not behave as we wish, so we apply feedback to transform it into a system that does.

Suppose that we take a closer look at a motor position controller, with a potentiometer to measure the output position and some other constants defined for good measure.

As usual we have state variables x and v, representing the position and velocity of the output.

The 'realistic' motor has a top speed; it does not accelerate indefinitely. Let us define the input, u, in terms of the proportion of full drive that is applied. If $u = 1$, the drive amplifier applies the full supply voltage to the motor.

Now when we apply full drive, the acceleration decreases as the speed increases. Let us say that the acceleration is reduced by an amount $a.v$.

If the acceleration from rest is b for a maximum value 1 of input, then we see that the velocity is described by the first-order differential equation

$$\dot{v} = -av + bu$$

DOI: 10.1201/9781003363316-11

The position is still a simple integral of v,

$$\dot{x} = v$$

If we are comfortable with matrix notation, we can combine these two equations into what appears to be a single equation concerning a vector that has components x and v:

$$\begin{bmatrix} \dot{x} \\ \dot{v} \end{bmatrix} = \begin{bmatrix} 0 & 1 \\ 0 & -a \end{bmatrix} \begin{bmatrix} x \\ v \end{bmatrix} + \begin{bmatrix} 0 \\ b \end{bmatrix} u$$

As shorthand, we can represent the vector as $\mathbf{x}$ and the input as $\mathbf{u}$ (this will allow us to have more than one component of input), so the equation appears as:

$$\dot{\mathbf{x}} = \begin{bmatrix} 0 & 1 \\ 0 & -a \end{bmatrix} \mathbf{x} + \begin{bmatrix} 0 \\ b \end{bmatrix} \mathbf{u}$$

This is an example of our typical form

$$\dot{\mathbf{x}} = \mathbf{A}\mathbf{x} + \mathbf{B}\mathbf{u}$$

Exercise 11.1
What is the top speed of this motor?

Exercise 11.2
What is the time constant of the speed's response to a change in input?

Now for feedback purposes, what concerns us is the output voltage of the potentiometer, y. In this simple case, we can write $y = cx$, but to be more general we should regard this as a special case of a matrix equation:

$$\mathbf{y} = \mathbf{C}\mathbf{x}$$

In this example, we can only measure the position. We may be able to guess at the velocity, but without adding extra filtering circuitry we cannot feed it back. If on the other hand we had a tacho to measure the velocity directly, then the output y would become a vector with two components, one proportional to position and the other proportional to the velocity.

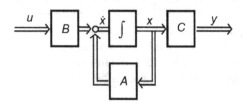

State equations in block diagram form.

For that matter, we could add a sensor to measure the motor current, and add a few more sensors into the bargain. They would be of little help in controlling this particular system, but the point is that the number of outputs is simply the number of sensors. This number might be none (not much hope for control there!), or any number that could even be more than the number of state variables. Adding extra sensors has a useful purpose when making a system such as an autopilot, where the control system must be able to survive the loss of one or more signals.

11.2 FEEDBACK

The input to our system is at present the vector **u**, which here has only one component. To apply feedback, we must mix proportions of our output signals with a command input **w** to construct the input **u** that we apply to the system.

To understand how the 'command input' is different from the input **u**, think of the cruise control of a car. The input to the engine is the accelerator. This will cause the speed to increase or decrease, but without feedback to control the accelerator the speed does not settle to any particular value.

The command input is the speed setting of the cruise control. The accelerator is now automatically varied according to the error between the setting and the actual speed, so that the car responds to a change in speed setting with a stable change to the new set speed.

Now by mixing sensor values **y** and command inputs **w** to make the new input, we will have

$$\mathbf{u} = \mathbf{Fy} + \mathbf{Gw}$$

When we substitute this into the state equation, we obtain:

$$\dot{\mathbf{x}} = \mathbf{Ax} + \mathbf{B}(\mathbf{Fy} + \mathbf{Gw})$$

But we also know that

$$\mathbf{y} = \mathbf{Cx}$$

so

$$\dot{\mathbf{x}} = \mathbf{Ax} + \mathbf{B}(\mathbf{FCx} + \mathbf{Gw})$$

i.e.

$$\dot{\mathbf{x}} = (\mathbf{A} + \mathbf{BFC})\mathbf{x} + \mathbf{BGw}$$

So by adding this feedback, the system has been changed to have a new set of equations, in which the matrix **A** has been replaced by (**A** + **BFC**) and where a new matrix **BG** has replaced the input matrix **B**.

As we saw at the end of chapter ten, our task in designing the feedback is to choose the coefficients of the feedback matrix **F** to make (**A** + **BFC**) represent something with the response we are looking for.

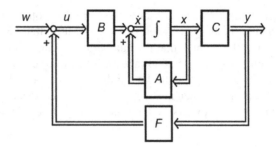

Feedback in block diagram form.

If the input and output matrices, **B** and **C**, both had four useful elements then we could choose the four components of **F** to achieve any closed-loop state matrix we wished. Unfortunately, in this case there is a single input and a single output, the position signal, so the **B** and **C** matrices degenerate to:

$$\mathbf{B} = \begin{bmatrix} 0 \\ b \end{bmatrix}$$

and

$$\mathbf{C} = \begin{bmatrix} c & 0 \end{bmatrix}$$

so that **F** has reduced to a single scalar f. We see that:

$$\mathbf{BFC} = \begin{bmatrix} 0 & 0 \\ b \cdot f \cdot c & 0 \end{bmatrix}$$

so that

$$\mathbf{A} + \mathbf{BFC} = \begin{bmatrix} 0 & 1 \\ bfc & -a \end{bmatrix}$$

To make it clear what this means, let us look at the problem using familiar techniques.

11.3   A SIMPLER APPROACH

We can attack the position control example in the "traditional" way as follows. We found a differential equation for the motor speed

$$\dot{v} = -av + bu$$

We can use the relationship between velocity v and position x to see it as a second-order equation

$$\ddot{x} = -a\dot{x} + bu \tag{11.1}$$

Now the potentiometer gives an output y proportional to x, so

$$y = c\,x$$

Feedback means mixing this potentiometer signal with another input w, the target position. We apply the result to the input so that

$$u = f\,y + g\,w$$

Substituting this into equation 11.1 we get:

$$\ddot{x} = -a\dot{x} + b\big(fcx + gv\big)$$

or

$$\ddot{x} + a\,\dot{x} - b\,f\,c\,\dot{x} = b\,g\,w \tag{11.2}$$

It is in the classical form of a second-order differential equation, showing a variable and its derivatives on the left and an input on the right.

Let us give some values to the constants.

Suppose that the motor has a time constant of a fifth of a second, so that $a = 5$.

If full drive will provide an acceleration from standstill of six units per second per second, then $b = 6$.

Finally, let the position transducer give us $c = 1$ volt per unit. Suppose also that we have chosen to set the feedback coefficient $f = -1$ and the command gain to be $g = 1$. Then equation 11.2 becomes

$$\ddot{x} + 5\dot{x} + 6x = 6w \tag{11.3}$$

If w is constant, this has a 'particular integral' where $x = w$, meaning that x is settled at the target.

To complete the general solution, we need to find the 'complementary function', the solution of

$$\ddot{x} + 5\dot{x} + 6x = 0 \tag{11.4}$$

Suppose that the solution is of the form

$$x = e^{mt}$$

Then

$$\dot{x} = me^{mt}$$

and

$$\ddot{x} = m^2 e^{mt}$$

When we substitute these into equation 11.4, we get

$$m^2 e^{mt} + 5m e^{mt} + 6e^{mt} = 0$$

and we can take out the exponential as a factor to get

$$\left(m^2 + 5m + 6\right) e^{mt} = 0$$

Since the exponential will not be zero unless mt has a value of minus infinity, we have

$$m^2 + 5m + 6 = 0$$

and we can solve for the value of m, to find

$$m = -2 \text{ or } m = -3$$

The complementary function is a mixture of two exponentials

$$x = A e^{-2t} + B e^{-3t}$$

where A and B are constants, chosen to make the function fit the two initial conditions of position and velocity.

Later chapters will delve into this more deeply.

12 Putting It into Practice

ABSTRACT

A practical experiment gives the incentive for some deeper analysis. A simple kit can be bought in the hobbyist market. With the aid of a BBC Micro:bit, Arduino or other microcontroller, it can be turned into a challenging balancing problem. To the state equations and simulation can be added the technique of 'pole assignment', as a way to choose those tricky feedback parameters.

12.1 INTRODUCTION

A practical control task is an essential part of any course in control. There is a vast range of choices for the design of such a system. Something must move, and an important decision must be made on the measurement of that movement. Whether linear or rotary, the measurements can be split into relative and absolute, into continuous or discrete.

An inverted pendulum has been a long-term favourite as a laboratory experiment. It is relatively easy to construct and by no means as difficult to control as the vendors of laboratory experiments would have you believe. Such a system has the added advantage that the pendulum can be removed to leave a practical position control system, of the sort that we have already considered.

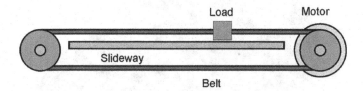

A position control system.

With a small motor connected to the shaft of the drive motor, a velocity signal can be made available. A ten-turn potentiometer can be connected to the shaft at the idler end, to give an analogue position signal. These voltages are applied to the analogue inputs of a microprocessor such as an Arduino, which can use an H-bridge driver to do the rest.

With such a system, important features of position control can be explored that are so often hidden in laboratory experiments. Although settling time and final accuracy are important, for the design of an industrial controller it is necessary to test its ability to withstand a disturbing force. Bearing all safety measures in mind, it is important for the experimenter to test the stiffness by giving the load a push.

DOI: 10.1201/9781003363316-12

Now a stick for the pendulum can be mounted on the load, with its angle of tilt measured by that Hall effect sine-cos sensor or a commercial equivalent.

We have already gone a little way towards exploring the mathematics of such a system, by looking at the task of riding a bicycle. You have also met a Segway simulation, presented merely as an example of the 'move-the-blocks' visualisation method. Now is the time to look more deeply into an easily constructed laboratory experiment.

To make an inverted pendulum experiment would probably involve the efforts of a workshop. But there is an alternative that demonstrates the same principles, which can be built as a DIY project with components from a hobby store.

12.2   A BALANCING TROLLEY

This task of balancing an inverted load is really a Segway in disguise. It is similar to that of an inverted pendulum, but construction is much simpler. Components can be bought cheaply from a hobby site to build a two-wheeled trolley. This would normally have a skid to serve as a third wheel, but this can be removed, adding a weight to put the centre of mass above the wheels.

A balancing trolley.

The chassis is a simple hobby kit, which comes with a skid that has been removed. Its control is based on a microcomputer card, the BBC Micro:bit. This is seen at the top of the picture. It is plugged into a 'Motor:bit' driver card to complete the driving of the two wheels.

An alternative control card would be an Arduino, there are many choices. However, the Micro:bit comes with built-in acceleration and magnetic sensors. It is an added challenge to achieve balancing without adding any more.

That sums up the construction, but the essence of the task is the design of the control system. We must generate a set of state equations to describe the system.

Clearly there are four state variables, the position and velocity of the wheels and the tilt and tilt rate of the cart. We will denote these as x, v, θ and ω. The input is the force exerted by the motor through the wheels.

Two of the differential equations are trivial,

$$dx/dt = v$$

and

$$d\theta/dt = \omega$$

but we have to think harder about the others.

The wheel force will both accelerate the mass and cause the rotation to accelerate giving us more equations to sort out. But that wheel force will not just be proportional to u. It will also be influenced by the motor speed and the inertia of the motor.

In fact, if the motor is reasonably highly geared, it is that motor speed that will dominate. State variables are not unique, provided they completely define the motion of the system. We can just as readily choose the tilt angle and the angle turned by the motor as state variables. But to avoid being caught up in gearbox matters, let us define φ as the angle turned by the wheels, relative to the trolley.

With our assumption that motor inertia dominates the dynamics, we can write

$$\ddot{\varphi} = au - b\dot{\varphi}$$

where a and b are constants that we must define. The time constant of the motor might be one quarter of a second, so that would define b as having a value of 4. The top speed of the motor is a/b radians per second, so if we assume that at the wheels it is about four revolutions per second, i.e. 25 radians per second, that would give a a value of 100. Putting in these values, we have

$$\ddot{\varphi} = 100u - 4\dot{\varphi}$$

But we have not yet defined a symbol for $\dot{\varphi}$. Let us choose its capital, Φ.

We can split this into two state equations

$$\dot{\varphi} = \Phi$$

and

$$\Phi = 100u - 4\Phi$$

Bearing the tilt in mind, the wheel will have rotated through $\theta + \varphi$ radians, so

$$x = r.(\theta + \varphi)$$

where r is the radius of the wheel.

But now we have to consider the dynamics of the tilt.

The acceleration component due to gravity is proportional to the height h of the centre of mass above the axle. But there is another component due to the linear acceleration.

So

$$\dot{\omega} = \frac{gh}{J}\theta - \frac{\ddot{x}}{h}$$

where J is the moment of inertia of the trolley and g is the gravitational acceleration.

We must convert that $\ddot{x}$ to our state variables, using $x = r(\theta + \varphi)$.

$$\dot{\omega} = \frac{gh}{J}\theta - \frac{r}{h}(\dot{\omega} + \dot{\Phi})$$

When we substitute for $\dot{\Phi}$ we get

$$\dot{\omega} = \frac{gh}{J}\theta - \frac{r}{h}(\dot{\omega} + 100u - 4\Phi)$$

i.e.

$$\left(1 + \frac{r}{h}\right)\dot{\omega} = \frac{gh}{J}\theta + 4\frac{r}{h}\Phi - 100\frac{r}{h}u$$

Let us try to put some values to the numbers. For simplicity let us assume that $r = h$. That simplifies the equation to

$$\dot{\omega} = \frac{gh}{2J}\theta + 2\Phi - 50u$$

We still have that annoying $gh/2J$ to worry about. But if we hang the trolley upside down, rotating freely about the axle, the equation of its pendulum swinging will be

$$\ddot{\theta} = -\frac{gh}{J}\theta$$

If it swings at two cycles per second, that gives a value of $2^2(2\pi)^2 = 158$ to gh/J, let us round it to 160.

So we now have our four equations:

$$\dot{\theta} = \omega$$

$$\dot{\omega} = 80\,\theta + 2\,\Phi - 50\,u$$

$$\dot{\varphi} = \Phi$$

$$\dot{\Phi} = 100u - 4\Phi$$

In matrix form the equations become

$$\begin{bmatrix} \dot{\theta} \\ \dot{\omega} \\ \dot{\varphi} \\ \dot{\Phi} \end{bmatrix} = \begin{bmatrix} 0 & 1 & 0 & 0 \\ 80 & 0 & 0 & 2 \\ 0 & 0 & 0 & 1 \\ 0 & 0 & 0 & -4 \end{bmatrix} \begin{bmatrix} \theta \\ \omega \\ \varphi \\ \Phi \end{bmatrix} + \begin{bmatrix} 0 \\ -50 \\ 0 \\ 100 \end{bmatrix} u$$

where the dots over the variables mean 'rate of change'.
   The code to simulate this would be:

```
theta=theta+omega*dt;
omega=omega+(80*theta+2*phirate-50*u)*dt;
phi=phi+phirate*
dtphirate=phirate+(-4*phirate+100*u)*dt
x=(theta+phi)*r;
v=(omega+phirate)*r;
```

We could instead express this with state variables x and v with a little substitution.

$$x = r(\theta + \varphi)x$$

$$v = r(\omega + \Phi)$$

Taking $r = 2$, we have $\varphi = x/2 - \theta$ and $\Phi = v/2 - \omega$.
   So

$$\dot{\omega} = 80\omega + 2(v/2 - \omega) - 50u$$

$$\dot{\omega} = 78\omega + 1v - 50u$$

To calculate dv/dt:

$$\dot{v} = 2\dot{\omega} + 2\dot{\mu}$$

$$\dot{v} = 160\,\theta + 4\Phi - 100u - 8\Phi + 200u$$

$$\dot{v} = 160\,\theta - 4\Phi + 100u$$

$$\dot{v} = 160\,\theta - 4(v/2 - \omega) + 100u$$

$$\dot{v} = 160\,\theta + 4\omega - 2v + 100u$$

This gives

$$\begin{bmatrix} \dot{\theta} \\ \dot{\omega} \\ \dot{x} \\ \dot{v} \end{bmatrix} = \begin{bmatrix} 0 & 1 & 0 & 0 \\ 78 & 0 & 0 & 1 \\ 0 & 0 & 0 & 1 \\ 160 & 4 & 0 & -2 \end{bmatrix} \begin{bmatrix} \theta \\ \omega \\ x \\ v \end{bmatrix} + \begin{bmatrix} 0 \\ -50 \\ 0 \\ 100 \end{bmatrix} u$$

or rearranging the order of the variables

$$
\begin{bmatrix} \dot{x} \\ \dot{v} \\ \dot{\theta} \\ \dot{\omega} \end{bmatrix} = \begin{bmatrix} 0 & 1 & 0 & 0 \\ 0 & -2 & 160 & 4 \\ 0 & 0 & 0 & 1 \\ 0 & 1 & 78 & 0 \end{bmatrix} \begin{bmatrix} x \\ v \\ \theta \\ \omega \end{bmatrix} + \begin{bmatrix} 0 \\ 100 \\ 0 \\ -50 \end{bmatrix} u
$$

Having defined the problem, now we must seek a solution.

As in the case of the bicycle, the first priority is to remain upright.

We need to feed back enough θ to overcome the destabilising term, plus enough ω to give good damping. Since we have not yet got a trolley, we must resort to a simulation at **Trolley1**.

We assume that all the states are available for feedback, so thinking of position and tilt we try

```
u=kx*x+kv*v+ktheta*theta+komega*omega;
```

Because of the large coefficients, it might be necessary to reduce the step length to .001 seconds.

Exercise 12.1
Run the code at **Trolley1** to find values that will stabilise the system. Can you improve on the default feedback coefficients of 0.5, 0.3, 20, 2, which were arbitrarily chosen?

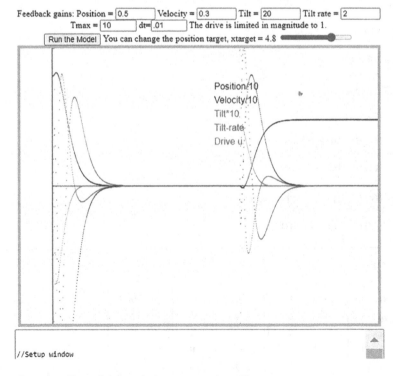

Screenshot of Trolley1.xhtml with the suggested coefficients.

12.3   GETTING MATHEMATICAL

Finding coefficients to stabilise the system is only half of the battle. Our control possibilities are going to be limited by the variables that we can sense. The Micro:bit has a built-in accelerometer that will have a horizontal component if the trolley is tipped, but as well as $g \sin(\theta)$ it will add the trolley horizontal acceleration. Does this matter? Do we need a rate gyro chip? Do we need wheel transducers to measure position?

How would we measure x or v? They would present problems. But we can easily measure phi, the angles turned by the wheels relative to the body. So instead of feeding back x and v, let us consider feeding back phi and phirate.

The simulation at **Trolley2** lets you choose those gains instead.

Exercise 12.2
Load and experiment with **Trolley2** to try to get the 'best' performance.

But before considering the problems of transducers, there is the question of why the coefficients are all positive. For tilt it is obvious enough, but for position control we usually have to use negative feedback.

Remember the bicycle. Imagine that we are to the right of the target line. To turn to the left we had to lean to the left. To start to lean, we have to give the handlebars a flick to the right, before the lean feedback turns it to the left.

If you do not find that convincing, mathematics can be brought into play.

If we make u a combination of the variables

$$u = a * \theta + b * \omega + c * \varphi + d * \Phi$$

we can write this in a more formal way as

$$\begin{bmatrix} a & b & c & d \end{bmatrix} \begin{bmatrix} \theta \\ \omega \\ \varphi \\ \Phi \end{bmatrix}$$

When we substitute for u in the matrix equation we get

$$\begin{bmatrix} \dot{\theta} \\ \dot{\omega} \\ \dot{\varphi} \\ \dot{\Phi} \end{bmatrix} = \begin{bmatrix} 0 & 1 & 0 & 0 \\ 80 & 0 & 0 & 2 \\ 0 & 0 & 0 & 1 \\ 0 & 0 & 0 & -4 \end{bmatrix} \begin{bmatrix} \theta \\ \omega \\ \varphi \\ \Phi \end{bmatrix} + \begin{bmatrix} 0 \\ -50 \\ 0 \\ 100 \end{bmatrix} \begin{bmatrix} a & b & c & d \end{bmatrix} \begin{bmatrix} \theta \\ \omega \\ \varphi \\ \Phi \end{bmatrix}$$

i.e.

$$\begin{bmatrix} \dot{\theta} \\ \dot{\omega} \\ \dot{\varphi} \\ \dot{\Phi} \end{bmatrix} = \begin{bmatrix} 0 & 1 & 0 & 0 \\ 80-50a & -50b & -50c & 2-50d \\ 0 & 0 & 0 & 1 \\ 100a & 100b & 100c & -4+100d \end{bmatrix} \begin{bmatrix} \theta \\ \omega \\ \varphi \\ \Phi \end{bmatrix}$$

The magic lies in equating to zero the determinant of

$$\begin{bmatrix} 0-\lambda & 1 & 0 & 0 \\ 80-50a & -50b-\lambda & -50c & 2-50d \\ 0 & 0 & 0-\lambda & 1 \\ 100a & 100b & 100c & -4+100d-\lambda \end{bmatrix}$$

Now, recall that we can add a multiple of one row to another, without changing the value of the determinant. So, we will add half the bottom row to the second row to get

$$\begin{bmatrix} -\lambda & 1 & 0 & 0 \\ 80 & -\lambda & 0 & \dfrac{-\lambda}{2} \\ 0 & 0 & -\lambda & 1 \\ 100a & 100b & 100c & -4+100d-\lambda \end{bmatrix}$$

We would normally be tempted to expand by the rows with most zeros, but instead we will expand by the bottom row.

We get:

$$-(100a)\left(-\lambda^2/2\right)$$

$$+(100b)(-\lambda)\left(-\lambda^2/2\right)$$

$$-(100c)\left(\lambda^2-80\right)$$

$$+(-4+100d-\lambda)\left(-\lambda^3+80\lambda\right)$$

Gathering powers of lambda:

$$\lambda^4+(4-100d+50b)\lambda^3+(50a-100c-80)\lambda^2+(-320+8000d)\lambda+8000c$$

When this is equated to zero, the roots define the response.

For reasons that will be explained much later, all the coefficients of

$$\lambda^4+(4-100d+50b)\lambda^3+(50a-100c-80)\lambda^2+(-320+8000d)\lambda+8000c=0$$

must have negative real parts. One of the conditions for that to be true is that all the coefficients of the equation must be positive. We can make the following deductions:

- a and b, the coefficients of tilt and tilt rate feedback, must both be positive.
- c and d must both be positive – they closely represent x and v.

Can we deduce more than this?

12.4   POLE ASSIGNMENT

The solutions of the equation

$$\lambda^4 + (4 - 100d + 50b)\lambda^3 + (50a - 100c - 80)\lambda^2 + (-320 + 8000d)\lambda + 8000c = 0$$

are called its roots or poles. In the position control example, we saw that the zero-input response is a function that includes the exponentials of these roots, thus we want to be sure that the exponentials are stable.

Rather than merely looking for roots that are stable, we can choose feedback coefficients that will give us a performance that we specify. If we are too ambitious, the numbers could be unreasonable. So, what coefficients should we look for?

We can associate the roots with the separate tasks of correcting a position error and correcting the tilt. We would not want to correct the position in much under two seconds, so a quadratic equation representing critical damping for this part might be

$$\lambda^2 + \lambda + 0.25 = 0$$

To correct the tilt, we would look for a time constant of the same order as the instability, so the equation

$$\lambda^2 + 32\lambda + 250 = 0$$

could seem appropriate. If we multiply these together, we get

$$\lambda^4 + 33\lambda^3 + 282.25\lambda^2 + 258\lambda + 62.5 = 0$$

When we compare this with the equation and equate coefficients, we get equations for the coefficients:

$$4 - 100d + 50b = 33$$

$$50a - 100c - 80 = 282.25$$

$$-320 + 8000d = 258$$

$$8000c = 62.5$$

Exercise 12.3
Calculate and try these values in **Trolley2**.

Exercise 12.4
Maybe the position time constant needs to be reduced to one second. Calculate new feedback values and try these.

The one-second time constant would give us the product of quadratics

$$(\lambda^2 + 2\lambda + 1)(\lambda^2 + 32\lambda + 250) = 0$$

which expands to

$$\lambda^4 + 34\lambda^3 + 314\lambda^2 + 532\lambda + 250 = 0$$

Compare it again with

$$\lambda^4 + (4 - 100d + 50b)\lambda^3 + (50a - 100c - 80)\lambda^2 + (-320 + 8000d)\lambda + 8000c = 0$$

and get a new set of gains. Try those values.

13 Observers

ABSTRACT

What can you do if your control system is lacking an essential velocity signal? You construct one from the position signal, using a 'filter' as part of an 'observer'. We look deeper into the trolley problem, where the main task is to stay upright. Can sufficient be gleaned from the accelerometer contained in the BBC Micro:bit to achieve the task?

13.1 INTRODUCTION

This carries on the topic of the previous section, the implementation of a practical control system. We have looked at the selection of feedback parameters if all state variables can be measured, but now must decide what to do with the sensor signals that we actually possess.

In module eight, we saw that it was possible to estimate a derivative by taking differences, either subtracting a previous value or a variable that has been 'slowed'.

13.2 LAPLACE AND HEAVISIDE

The concept of the Heaviside 'operator' is very simple. Instead of writing d/dt, you just write D.

The equation

$$d^2x/dt^2 + 5dx/dt + 6x = u$$

becomes

$$\left(D^2 + 5D + 6\right)x = u$$

And we will be tempted to write

$$x = \frac{1}{D^2 + 5D + 6}u$$

But what does this mean?

The fact that D is an operator, not a variable, is complicated by the fact that the solution of

$$Dx = 0$$

DOI: 10.1201/9781003363316-13

is not zero, but any constant initial condition.

The great attraction of the Laplace transform was that instead of an operator D, you could think in terms of a variable, s. But it came with a bag and baggage of relationships that gave no real advantages over D. There were indeed some disadvantages.

Both found their main use in the ability to look up an already-known solution to a differential equation, but Laplace picks up an extra s. For Heaviside, the function '1' is the unit step, zero for all history and unity after $t=0$. For Laplace it is the unit impulse, zero everywhere except for an infinite spike of zero width at $t=0$. It is the integral of this spike that is the Heaviside unit step.

Both D and s are used in exactly the same way. The input u is replaced by its transform equivalent and the resulting expression is applied to a look-up table. If there is a polynomial in the denominator, it is factorised and put into partial fractions, fragments that are more likely to appear in that table.

13.3 FILTERS

When we apply some dynamics to a signal, we are 'filtering' it. It does not matter if we use D or s, because we are looking at a 'transfer function', not a time solution.

So we can think of v as sx, the derivative of x. But a true derivative is something very hard to obtain.

If we apply the filter '1' to a unit step, we still have that same step. But what happens when we apply $1/(1 + \tau s)$ to it?

We are looking for the solution to the equation

$$x + \tau\, dx/dt = 1$$

when x starts at zero. It is not hard to find the solution

$$x = 1 - e^{-t/\tau}$$

The filter has applied a 'lag' to the signal. The time constant in this case is τ. The response is shown in Figure 13.1.

Now consider what we get if we subtract this from the original step.

$$1 - \frac{1}{1+\tau s} = \frac{1+\tau s}{1+\tau s} - \frac{1}{1+\tau s}$$

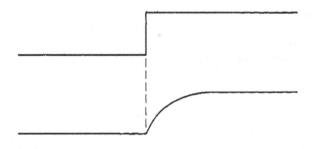

FIGURE 13.1 Effect of applying a lag.

giving

$$\frac{\tau s}{1 + \tau s}$$

We get a response like that in Figure 13.2.

This is something that we can do with analogue circuitry. Although it is a lagged derivative, it serves well when added to the original signal as a 'phase advance'. Later we will see how a simple circuit can stabilise a second-order system.

13.4   THE KALMAN FILTER

For half a century, the Kalman filter has been regarded with awe. But the idea behind it is very simple.

Suppose that you have an accurate model of the system that you are trying to control. If you apply the same input signals to both of them, will they not have the same internal states? Of course, the answer is 'no'. For the two-integrator motor the initial states might not be the same. The states of the system and model will drift apart.

But if we take the difference between the outputs of system and model, we can feed this back anywhere in the model to pull it into line. Take that motor example:

Figure 13.3 shows that whereas we can only apply a feedback signal to the first integrator of the motor, we are free to choose p and q to apply feedback to both integrators of the model. Not only do we have both of the estimated states available, measurement noise on the output can be reduced in the estimated version.

The disadvantage is that the resulting control filter might be much more complicated than we really need. A simple phase advance might be sufficient.

13.5   THE BALANCING TROLLEY EXAMPLE

Our problem here is of a different nature. We already have an accelerometer signal. Is this enough for control, or do we have to add other sensors?

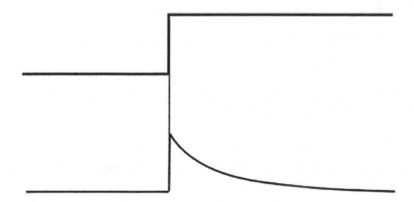

FIGURE 13.2   The resulting high-pass filter.

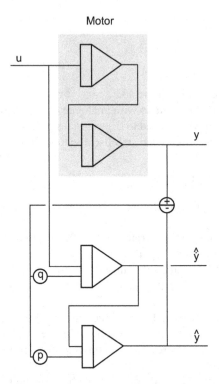

FIGURE 13.3   A simple example of a Kalman filter.

Part of the accelerometer output will be $g\theta$, and a phase advance might be enough to estimate the necessary ω component. But the signal will also be influenced by the linear acceleration of the sensor due to the drive. Will this make the signal unusable by feeding a component of drive straight back to the drive?

Let us look at those state equations again, simplified to ignore motor damping. For the acceleration at the axle, we had

$$\ddot{x} = 200u$$

and for the tilt we had

$$\ddot{\theta} = 80\,\theta - 50u$$

On the body of the trolley, at a height a above the axle, we will have acceleration

$$\ddot{x} + a\ddot{\theta}$$

So, if we put the sensor 4 cm above the axle, the u terms should cancel out. The sensor should measure an acceleration

$$-g\theta$$

It might even work! But there is a simple way to find out, build the trolley and test it!

The sensor has to be put at the 'centre of percussion' relative to the axle, and you can find that centre experimentally. Attach a 'mast' to the trolley, a vertical stick above the axle, maybe some 10 cm long.

Suspend the trolley upside down as you did before, letting it rotate freely about the axle. I simply used a loop of string about the axle. Now shake the axle fore and aft, while watching the mast. You should see a point on the mast that is relatively undisturbed. That is the centre of percussion, where you should mount your sensor.

But if that fails, we could add a gyro sensor, something that will be unaffected by acceleration.

13.6   COMPLEMENTARY FILTERING

In systems such as autopilots, there are frequently sensors that seem redundant. An example would be a bank angle sensor and a roll rate gyro. But by combining the signals with a filter, shortcomings can be removed.

If we have a sensor for y and another sensor for $\dot{y}$, they might both have a datum error. If we add them to make $y/\tau + \dot{y}$ we can apply a high-pass filter

$$\frac{\tau s}{1+\tau s} \quad \frac{(1+\tau s)y}{\tau}$$

to remove any datum. The result will be a clean derivative sy with no datum.

If we find that we have to use a rate gyro as well as the accelerometer, can we use a similar method to calculate θ from it? If we simply integrate it, our integral will drift off. Maybe we can apply a low-pass filter to it, to avoid this, after combining it with the accelerometer signal in some way.

The problem with the accelerometer signal was that unless its location was precisely balanced at that centre of percussion, it would feed the motor drive straight back to itself. This would result in either locking up or reducing the gain.

Let us see if we can use the accelerometer alone. It does not matter whether we can construct a tilt angle signal, we just have to feed back something that will stabilise the system.

Without ploughing through the mathematics, we can try it on a simulation.

13.7   A PRAGMATIC APPROACH

The simulation **Trolley3** investigates the effect of putting the sensor at the wrong height. The default height in this simulation is two centimetres above the axle. You can edit this to try different values. You will see that getting the right height is essential.

Unless we add sensors to the wheels, however, we will have to guess at the velocity and distance. If we do not need accuracy, we can model the motor in the controller by lagging the drive signal to estimate the motor's speed. This has been done in **Trolley4**.

Exercise 13.1
Run **Trolley4**. Try to improve on the default values.

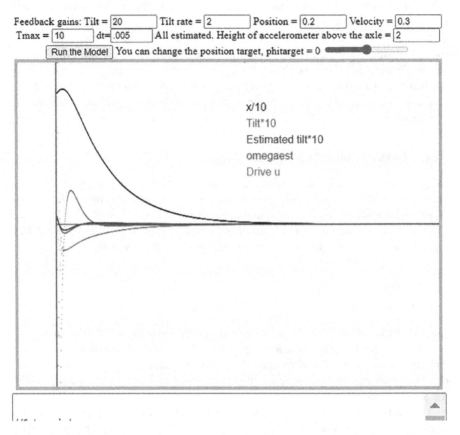

Feedback gains: Tilt = 20 | Tilt rate = 2 | Position = 0.2 | Velocity = 0.3
Tmax = 10 | dt= .005 | All estimated. Height of accelerometer above the axle = 2
Run the Model | You can change the position target, phitarget = 0

x/10
Tilt*10
Estimated tilt*10
omegaest
Drive u

Screen grab of Trolley4.xhtml with the default parameters.

If possible, apply all this to balancing a real physical trolley.

14 More about the Mathematics

ABSTRACT

Now, we must turn back to the mathematics and take our system by its roots. What are 'poles' and 'zeroes' and what do we mean by a complex exponential? They are not as forbidding as they seem. Imaginary exponentials are just sines and cosines in disguise.

14.1 INTRODUCTION

In the previous material, there was an effort to avoid any need for serious mathematics. There were hints of roots and higher-order differential equations, but the emphasis was on state equations and simulation. Now we are going to have to look more deeply at the relevance of complex roots, both for continuous and discrete time systems. We will also have to investigate vectors and matrices, seeing the significance of eigenvectors and eigenvalues.

But at the same time, the practical aspects will be borne in mind.

14.2 HOW DID THE EXPONENTIALS COME IN?

If we look for the solution of

$$\ddot{x} + 5\dot{x} + 6x = u$$

we can split the search into two parts.

First we look for a 'particular integral', some function that fits the equation. For example, if u is a simple constant of value 6, then a particular integral is $x = 1$, which is also constant.

But we must say 'a particular integral' because this might not be the only solution. To find other solutions, we must look for a 'complementary function' that satisfies

$$\ddot{x} + 5\dot{x} + 6x = 0$$

When we add this to the particular integral, it will have parameters that we can adjust to fit the answer to the initial conditions.

Let us consider an even simpler equation,

$$\dot{x} + ax = a$$

DOI: 10.1201/9781003363316-14

A particular integral is again a constant, $x = 1$.

For the complementary function, we solve

$$\dot{x} + ax = 0$$

If we want to be formal, we can rearrange this as

$$\frac{\dot{x}}{x} = -a$$

which we can integrate with respect to time to get

$$\log_e x = -at + b$$

We then take the exponential of both sides to give

$$x = e^{b-at}$$

$$x = Ae^{-at}$$

where the constant $A = e^b$.

But we can probably guess the solution without this formality.

So, our solution is

$$x = 1 + Ae^{-at}$$

Now to put a value to A, we must look at the initial conditions. If we have $x(0) = 0$, then $A = -1$ and

$$x = 1 - e^{-at}$$

14.3   MORE ABOUT ROOTS

Whenever we are dealing with linear, continuous systems we will meet polynomials. These could be in D, s, λ or m, depending only on which notation we choose. For the second-order equation

$$\ddot{x} + 5\dot{x} + 6x = u$$

We might say, "Let us try $x = Xe^{mt}$", which will find us looking for roots of

$$m^2 + 5m + 6 = 0$$

We might be using the Laplace notation of

$$X = \frac{1}{s^2 + 5s + 6} U$$

or trying to find eigenvalues of matrices by looking for the roots of their 'character-istic polynomial'. In every case we will be looking for roots, values which make the polynomial zero.

In this particular case, the roots are −2 and −3.

That means that our solution will include a complementary function consisting of multiples of e^{-2t} and e^{-3t}, plus terms contributed by u.

The conventional teaching material on Laplace transforms would then go into details of expressing the equations as partial fractions, so that time solutions could be looked up from a table. But that is not what really interests us when designing control systems. Our concern is whether the controlled system is stable, and how quickly any disturbances will die away.

In this case, we can see that the e^{-2t} contribution will decay to $1/e$ of its value in half a second. The e^{-3t} component will decay even faster, so we have no worries about stability. Real roots are no worry, provided they are negative, and the more negative the better.

14.4 IMAGINARY ROOTS

Now consider the system

$$\ddot{x} + x = 0$$

We are looking for something that when differentiated twice will give the negative of the same function.

If we assume a solution of the form e^{mt}, we would be looking for the roots of

$$m^2 + 1 = 0$$

giving

$$m = +/- j$$

What does it mean when we take the exponential of an imaginary number?

We also know that if we differentiate sin(t) or cosine(t) twice, they have that same property.

There are many ways to show that

$$e^{jt} = \cos t + j \sin t$$

One way is by inspection of a power series expansion. For now, just take it on trust.

We also have

$$e^{-jt} = \cos t - j \sin t$$

We can see that

$$\cos t = \frac{e^{jt} + e^{-jt}}{2}$$

and

$$\sin t = \frac{e^{jt} - e^{-jt}}{2}$$

The solution will of course be real, of the form

$A \cos t + B \sin t$

where A and B are determined by the initial conditions. It looks as though a lot of complex arithmetic is ahead, but there is an engineering solution.

$A \cos t + B \sin t$

is the real part of

$(A - jB)(\cos t + j\sin t)$

We do not really care what the imaginary part is, we can just work out our solution in the form of a complex number, $(A + jB)$, then multiply it by $(\cos t + j \sin t)$ and take the real part.

When u is a sinusoidal input, complex numbers help again.

Exercise 14.1
A system is defined by

$$\ddot{x} + 5\dot{x} + 6x = \sin t$$

What function defines x when the initial transient has died away?

We can start by observing that the input is the real part of $-je^{jt}$. We can deduce that x will also be a multiple of e^{jt}, so we can write x as the real part of Xe^{jt}, where X is a complex number.

$$X(j^2 + 5j + 6)e^{jt} = -je^{jt}$$

i.e.

$$X = \frac{-j}{-1 + 5j + 6}$$

We make the denominator real by multiplying top and bottom by $(5 - 5j)$ to get

$$X = -j(5 - 5j)/50$$

x is the real part of $(0.1 - 0.1j)(\cos t + j \sin t)$

$$x = 0.1\cos t + 0.1\sin t$$

Exercise 14.2

Solve this the 'conventional' way, starting with $x = A \cos t + B \sin t$. Which takes less effort?

14.5 COMPLEX ROOTS AND STABILITY

Let us look at an exponential of a complex number.

$$e^{(a+jb)t} = e^{at} e^{jbt} = e^{at} \cos bt + j e^{at} \sin bt$$

Multiply this by a complex constant and we have a mixture of $\sin bt$ and $\cos bt$, ruled by the exponential e^{at}.

Now if a is positive, the sine wave will grow exponentially. So for stability, the real part of every root must be negative. If it is zero, we have the knife-edge case where the sine wave goes on for ever.

When we looked at

$$\ddot{x} + x = 0$$

we saw a solution that involved $\sin t$ and $\cos t$.

More generally we would have an angular frequency ω, corresponding to f cycles per second, where $\omega = 2\pi f$.

$$\ddot{x} + \omega^2 x = 0$$

If we introduce a damping component, the roots can become complex. If we have

$$\ddot{x} + 2a\dot{x} + \omega^2 x = 0$$

we will have roots of

$$m^2 + 2am + \omega^2 = 0$$

giving

$$m = -a \pm \sqrt{a^2 - \omega^2}$$

As long as a is positive, we have no worries about stability.

If we start to increase a from zero, while it is less than ω, the roots will be complex.

Any disturbance will die away according to e^{-at}, which will be the envelope of damped oscillations. The angular frequency of those oscillations will start at ω, but decrease as a approaches ω.

When a reaches ω there is a significant change. First there will be two equal roots of $-\omega$, but as a increases further these will split. The larger one (in magnitude) will tend towards $-2a$, but the other one will decrease. Since the product of the roots is ω^2, as one increases the other root will decrease, so doubling a will tend to double the settling time. Remember that settling time is inversely proportional to the real part of the root.

Since the effect depends so strongly on the ratio between a and ω, the *damping factor* zeta is defined as

$$\zeta = a/\omega$$

Now the equation is written as

$$\ddot{x} + 2\zeta\omega\dot{x} + \omega^2 x = 0$$

When $\zeta = 1$, the system is 'critically damped'.

Exercise 14.3

Run the simulation at **Damping** to see the effect of increasing damping. Start with the default values and repeat with increments of zeta of 0.2. Especially note the response when $\zeta = 0.7$. See that as values are increased greater than one, the settling time gets longer.

Responses of $\ddot{x} + 2\zeta\omega\dot{x} + \omega^2 x = 0$

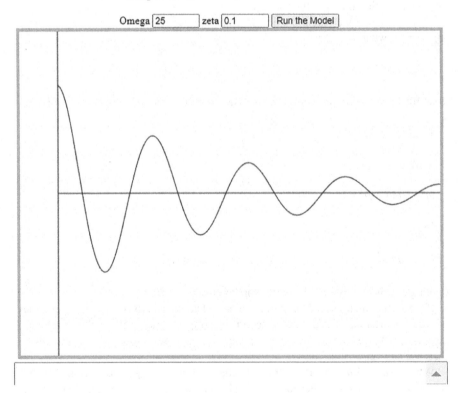

Omega 25   zeta 0.1   Run the Model

Screenshot of Damping.xhtml.

15 Transfer Functions

ABSTRACT

The last chapter was just an appetiser for more mathematics to come. By giving symbols such as D and s to the differential operator d/dt, we can make differential equations look like algebraic ones. If your concern is just to identify a function of time for the solution, you can manipulate the function to help you look up an answer from a table of known solutions. But that is not our aim here. We are interested in stability and the general way that the system responds to a disturbance. In fact, the system can have several inputs and outputs, leading us to a transfer-function matrix.

15.1 INTRODUCTION

Things are going to get a lot more mathematical from here, as we consider the transfer functions of systems and dynamic controllers.

We have been assuming that we know the details of the system that we are dealing with, but a large branch of control theory concerns exploring a system that is unknown. The traditional method was to use an oscillator and an attenuator. A sinusoidal signal was injected into the system and the *gain* was measured from the size of the signal that came out. As the frequency was varied, plots of gain against frequency gave insight into the transfer function.

When oscilloscopes came onto the scene, it became possible to measure the phase shift as well as the gain. The 'lag' that we have met obtained its name because the phase of the output sinewave would lag behind that of the input. We will see that to stabilise a second-order system, we need a *phase advance*.

When we describe our sinewaves with complex numbers, we have to be clear about how the numbers relate to lags and leads. We regard cos ωt as the 'benchmark', since it is the real part of $e^{j\omega t}$.

The sinewave sin ωt lags the cosine wave by 90 degrees. When the cosine wave is at its peak, the sinewave is just starting to increase.

So, cos ωt + sin ωt will have a 45 degrees lag, while cos ωt – sin ωt represents a phase lead. In complex number terms, $1 - j$ will denote a lag, while $1 + j$ will denote a lead.

There is another way to represent a complex number. That is by its *modulus* and *argument*.

The modulus of $a + jb$ is $\sqrt{a^2 + b^2}$, while the argument is $\tan^{-1} b/a$. So, you can interpret the modulus as the gain, and the argument as the amount of phase advance or lag.

A good place to start is with phase advance.

DOI: 10.1201/9781003363316-15

15.2   PHASE ADVANCE

When we considered a simple two-integrator system

$$\ddot{x} = u$$

we saw that feeding back x alone would result in oscillation – or in runaway instability if the coefficient is positive. We had to construct some sort of velocity signal to feed back with it.

We estimated the velocity *vest* by subtracting a lagged version of the position x

$$\frac{1}{1+\tau\,s}\,x$$

from x to get

$$\tau \text{vest} = \frac{\tau s}{1+\tau s}\,x$$

If we feed back a combination of these

$$u = -a\tau\ \text{vest} - bx$$

we will have

$$\ddot{x} + a\frac{\tau s}{1+\tau s}\,x + bx = 0$$

This really means that we are applying a phase-advance transfer function to the feedback

$$u = -\frac{b+[a+b]\tau s}{1+\tau s}\,x$$

So

$$\ddot{x} + \frac{b+[a+b]\tau s}{1+\tau s}\,x = 0$$

If we represent $\ddot{x}$ as s^2x and multiply through by $1 + \tau s$, we get

$$\left(\tau s^3 + s^2 + [a+b]\tau s + b\right)x = 0$$

This is a third-order equation, and for stability it is not enough just to have positive coefficients. So, what is the other criterion?

15.2.1   MATHEMATICAL ASIDE

A cubic polynomial can be factorised into the product of a linear term and a quadratic. We have

$$(s+p)(s^2+qs+r)=0$$

We can multiply this out to get

$$s^3+(p+q)s^2+(pq+r)s+pr=0$$

We know that p, q and r must all be positive for stability, but what does this tell us about our cubic coefficients?

If we take the product of the middle coefficients, we get

$$(p+q)(pq+r)=pr+p^2q+pq^2+qr$$

If p, q and r are all positive, this must be bigger than pr, the product of the outer ones. So to be sure that q is positive, the product of the middle pair of coefficients must be greater than the product of the outer ones.

Returning to our equation

$$(\tau s^3+s^2+[a+b]\tau s+b)x=0$$

we see that the product of the inner coefficients is $[a+b]\tau$, while that of the outer ones is just $b\tau$. Even a small amount of phase advance can stabilise the system, though there will be stricter conditions for achieving an acceptable response.

15.3   A TRANSFER-FUNCTION MATRIX

A set of linear state equations can be written in a concise mathematical form as

$$\dot{x}=Ax+Bu$$

with a set y of outputs given by

$$y=Cx+Du$$

x is the set of state variables that describe the state of the system at any instant, while y is the set of outputs that can be measured for feedback purposes. The matrix D represents the instantaneous effect that an input can have on an output, rather than passing via system variables. It is usually assumed to be zero, since instantaneous feedback around u would add a tricky complication.

With states in mind, taking D as zero, we see that applying feedback $u=Cy$ gives us a closed-loop equation

$$\dot{x}=(A+BFC)x$$

from which we can assess stability. The A matrix of the open-loop system has been replaced by a new value.

Returning to the original equations, now considering A to represent our controlled system, we have a number of command signals u and a set of outputs y that respond to those commands. We might wish to know the transfer function that links each output to those inputs.

Using 's' to represent the differential operator, the first equation can be written as

$$sx = Ax + Bu$$

$$[sI - A]x = Bu$$

where I is a unit matrix, hence

$$x = [sI - A]^{-1} Bu$$

and so

$$y = C[sI - A]^{-1} Bu$$

Each output will be represented as combinations of transfer functions operating on each of the inputs.

Fundamental to calculating the inverse of $[sI - A]$ will be the need to calculate its determinant, so once again we will be considering a polynomial in s.

This is called the *characteristic polynomial* of the matrix A. If we are considering transfer functions, it will be a polynomial in s. If we are solving matrix state equations it can be a polynomial with the same coefficients, but in λ.

Exercise 15.1

The system

$$\ddot{x} = u$$

can be expressed as two first-order equations

$$\dot{x} = v$$

$$\dot{v} = u$$

where both x and v can be measured.

A state vector $(x, v)'$ can be written as x, and an output vector as y.

Write the equations in the matrix form

$$\dot{x} = Ax + Bu$$

$$y = Cx$$

The system is controlled by feeding back

$$u = 6(d - x) - 5v$$

where d is a demanded position target.

In matrix terms, this would become

$$u = Fy + Gd$$

Fill in the matrices of the closed-loop expression

$$\dot{x} = (A + BFC)x + BGd$$

Then evaluate $(A + BFC)$ to complete the new closed-loop equations for x and y.

Solution

This is really a matter of juggling equations with which we are already familiar. Though it seems an unnecessary complication in this case, for more complicated systems it is a way to bring computing power to bear on the solution.

First we have

$$\dot{x} = \begin{bmatrix} 0 & 1 \\ 0 & 0 \end{bmatrix} x + \begin{bmatrix} 0 \\ 1 \end{bmatrix} u$$

and

$$y = \begin{bmatrix} 1 & 0 \\ 0 & 1 \end{bmatrix} x$$

Our feedback is

$$u = \begin{bmatrix} -6 & -5 \end{bmatrix} y + [6]d$$

So, the full equation would look like

$$\dot{x} = \left(\begin{bmatrix} 0 & 1 \\ 0 & 0 \end{bmatrix} + \begin{bmatrix} 0 \\ 1 \end{bmatrix} \begin{bmatrix} -6 & -5 \end{bmatrix} \begin{bmatrix} 1 & 0 \\ 0 & 1 \end{bmatrix} \right) x + \begin{bmatrix} 0 \\ 1 \end{bmatrix} [6]d$$

This simplifies to

$$\dot{x} = \begin{bmatrix} 0 & 1 \\ -6 & -5 \end{bmatrix} x + \begin{bmatrix} 0 \\ 6 \end{bmatrix} d$$

and

$$y = \begin{bmatrix} 1 & 0 \\ 0 & 1 \end{bmatrix} x$$

which we could have deduced with very much less effort.

Exercise 15.2

Now calculate the transfer-function matrix expressing y as a function of d.

Solution

We are looking for

$$y = C[sI - A]^{-1} Bu$$

where u is now just our position demand d and C is the unit matrix.

We have to invert $[sI - A]$ – which for the closed-loop matrix is

$$\begin{bmatrix} s & -1 \\ 6 & s+5 \end{bmatrix}$$

The inverse is

$$\begin{bmatrix} \dfrac{s+5}{s^2+5s+6} & \dfrac{1}{s^2+5s+6} \\ \dfrac{-6}{s^2+5s+6} & \dfrac{s}{s^2+5s+6} \end{bmatrix}$$

For Bu we have

$$\begin{bmatrix} 0 \\ 6 \end{bmatrix} d$$

So, we arrive at

$$y = \begin{bmatrix} \dfrac{s+5}{s^2+5s+6} & \dfrac{1}{s^2+5s+6} \\ \dfrac{-6}{s^2+5s+6} & \dfrac{s}{s^2+5s+6} \end{bmatrix} \begin{bmatrix} 0 \\ 6 \end{bmatrix} d$$

i.e.

$$y = \begin{bmatrix} \dfrac{6}{s^2+5s+6} \\ \dfrac{6s}{s^2+5s+6} \end{bmatrix} d$$

There is our transfer-function matrix.

16 Solving the State Equations

ABSTRACT

There is even more mathematics here, this time looking at the way that 'transformations' can be applied to our state matrices. These transformations can be seen to break down a second-order system into a pair of first-order systems which are independent, apart from a shared input.

A simple program illustrates the nature of an eigenvector, while a phase-plane plot shows how it relates to a solution.

16.1 INTRODUCTION

More complicated mathematics will creep in here, too. But it can give you another way to look at things.

As we apply matrices to the analysis of a control system, we see analogies with their use in 3D geometry. Sets of state vectors give us a measure of what is happening, but their choice is not unique. For example, combinations of position and velocity can be used, as long as those combinations can be unscrambled.

Consider our familiar system

$$\ddot{x} + 5\dot{x} + 6x = u$$

$$y = x$$

If we construct mixtures

$$w_1 = v + 2x$$

$$w_2 = v + 3x$$

we see that

$$\dot{w}_1 + 3w_1 = \ddot{x} + 5\dot{x} + 6x$$

$$\dot{w}_2 + 2w_2 = \ddot{x} + 5\dot{x} + 6x$$

So instead of our earlier equations in which x and v were tied together, we have two independent state equations

$$\dot{w}_1 + 3w_1 = u$$

$$\dot{w}_2 + 2w_2 = u$$

DOI: 10.1201/9781003363316-16

where

$$y = w_2 - w_1$$

We have applied a *transformation* to our state $[x, v]$ to represent it as $[w_1, w_2]$ with

$$\begin{bmatrix} w_1 \\ w_2 \end{bmatrix} = \begin{bmatrix} 2 & 1 \\ 3 & 1 \end{bmatrix} \begin{bmatrix} x \\ v \end{bmatrix}$$

Despite the mathematical acrobatics, there is a simple way to look at this.

When we use the conventional way to solve

$$\ddot{x} + 5\dot{x} + 6x = 0$$

we see that the answer is

$$Ae^{-2t} + Be^{-3t}$$

The two new variables just represent these parts of the solution separately.

16.2   VECTORS AND MORE

When we plot velocity against position, instead of plotting them both against time, the geometric connection becomes more obvious.

The matrix state equation

$$\dot{x} = Ax$$

expresses a rate-of-change vector $\dot{x}$ as a matrix product with the state x.

So here we have

$$\begin{bmatrix} \dot{x}_1 \\ \dot{x}_2 \end{bmatrix} = \begin{bmatrix} 0 & 1 \\ -6 & -5 \end{bmatrix} \begin{bmatrix} x_1 \\ x_2 \end{bmatrix}$$

where we write x_2 for v.

We can view this as a vector equation, giving a direction in which the plot of x_2 against x_1 will be extended at each step. If that direction is aligned with the past trace, the result will be a straight line.

Exercise 16.1
Run the simulation at **Phase4**. It is a phase-plane plot of the system response to a disturbance. You can click on the screen at any time to start a new plot. You should find that there are two sets of starting points from which the response is a straight line, one with slope −2 and the other with slope −3.

Exercise 16.2
Now change the position feedback coefficient to 4. You should find the linear responses much easier to find now, with slopes −1 and −4.

Exercise 16.3
Now change the position feedback coefficient to 10. Can you still find two straight lines?

The straight lines are only possible if there are real roots to the characteristic equation.

16.3 EIGENVECTORS

Those straight-line solutions are called the eigenvectors of the A-matrix.

When a matrix multiplies a vector, it usually changes its size and changes its direction. But if the vector is an eigenvector, the size might change but the direction remains the same.

Exercise 16.4
Run the simulation at **Eigen1**. As you move your mouse across the panel, you will see both the vector and its transform. When they are aligned, you have an eigenvector. You can change the coefficients to explore other matrices.

Matrix A is:

2	1
1	2

Values are:
x = 1.72
y = -0.64
l = 2.80
m = 0.44

Ratio is:
r = 1.544
Move the mouse over the square
to control the blue vector (x,y)'
The red vector will show
(l,m)' = A(x,y)'

When the two vectors are in line,
you have an eigenvector.

You can edit the matrix.

Screenshot of Eigen1.xhtml.

16.4   A GENERAL APPROACH

When the matrix multiplies an eigenvector, the resulting vector is in the same direction, but its value can have been changed. The amount by which it is multiplied is called the *eigenvalue*.

For considering stability, we are just looking for the solutions of

$$\dot{x} - Ax = 0$$

where that 0 is a vector that has zeroes for all its components.

If ζ is an eigenvector that satisfies

$$\dot{\xi} = \lambda \xi$$

We can write

$$\lambda I \xi - A \xi = 0$$

i.e.

$$(\lambda I - A)\xi = 0$$

For mathematical reasons, this new matrix must be '*singular*', meaning that its determinant must be zero.

Taking the determinant gives us a polynomial in λ, and each of the roots will be an eigenvalue.

So if stability is our only concern, there is no need to find the eigenvectors, we just need to test the eigenvalues.

- Construct the matrix $(\lambda I - A)$.
- Take its determinant.
- Inspect the coefficients of the polynomial in λ to ensure that all the roots have negative real parts.

We already know that all the coefficients must be positive, and that for a third-order polynomial the product of the inner terms must be greater than the outer ones, but what do we do if the order is higher?

In the late nineteenth century, Routh and Hurwitz devised a test for any order, involving the determinants of a succession of matrices. If you really need the details, a web search for 'Routh–Hurwitz' will lead you to them.

16.5   EQUAL ROOTS

There is a complication if roots are equal.

Consider the equation

$$\ddot{x} + 2\dot{x} + 1 = 0$$

We have two equal roots of minus one. But we need two terms in the solution to satisfy the initial conditions of position and velocity. Ae^{-t} is not enough.

If we try

$$x = te^{-t}$$

we have

$$\dot{x} = e^{-t} - te^{-t}$$

and

$$\ddot{x} = -e^{-t} - e^{-t} + te^{-t}$$

We see that te^{-t} also satisfies the equation, so our solution is

$$x = Ae^{-t} + Bte^{-t}$$

We have the same complication if we try to solve

$$\dot{x} + x = e^{-t}$$

And, it is solved the same way by

$$x = te^{-t} + Ae^{-t}$$

17 Discrete Time and the z Operator

ABSTRACT

So far, we have been regarding our digital simulation as an approximation to the continuous world. Now we must grit our teeth to look at systems in which the controller only applies feedback at intervals. We meet the z-transform, usually treated with awe as something derived from a Laplace transform. But we realise that it has a simple interpretation as a line of computer code. Instead of differential equations, we have difference equations. Instead of exponentials, we have solutions that are simple powers of a variable.

17.1 INTRODUCTION

We have met the D operator and the variable s. Now it is time to look at the discrete 'next' operator and the variable z.

These have a lot in common. Because of the difficulty of physically differentiating something, in a filter, s is most often found in the denominator, in the form of an integral. For a similar reason, z is often found as z^{-1}, meaning 'previous'.

But they can take their place on the left-hand side of an equation, so an integrator can be represented as

$$s\,x = u$$

or as a filter

$$x = \frac{1}{s}u$$

When time is discrete, with sampling at intervals T, the input is assumed to remain constant throughout the interval. So the integrator becomes a 'summer', and the next value of x is given by

$$\text{next}\,x = x + T\,u$$

We can use z to represent 'next' and write

$$z\,x = x + Tu$$

which we can rearrange as a discrete-time filter

$$x = \frac{1}{z-1}Tu$$

DOI: 10.1201/9781003363316-17

17.2   FORMAL METHODS

A mathematics textbook dealing with sequences will abound with subscripts, such as

$$x_{n+1} = x_n + x_{n-1}$$

Where we started our investigation of linear differential equations, we tried solutions of the form $x = Ae^{mt}$. With sequences we try solutions $x = Ak^n$

Once again, we have a polynomial to solve. Take the example

$$x_{n+1} = x_n + x_{n-1}$$

which we can write as

$$x_{n+2} - x_{n+1} - x_n = 0$$

If we try the $x = Ak^n$ solution, we get

$$k^2 - k - 1 = 0$$

This has roots $\left(1 \pm \sqrt{5}\right)/2$, i.e. 1.618 and –0.618.

Why is this worth mentioning?

That equation gives the famous Fibonacci sequence, in which each number is the sum of the two previous ones.

1, 1, 2, 3, 5, 8, 13, 21, 34, 55, 89, 144 and so on.

With a root greater than one, the series is heading off exponentially. What would the condition need to be for a sequence to converge? Are the roots always real?

Let us try a similar series, but with the following number halved:

$$x_{n+1} = \left(x_n + x_{n-1}\right)/2$$

Now our polynomial becomes

$$k^2 - 0.5k - 0.5 = 0$$

It has roots $k = 1$ and $k = -0.5$. So what kind of sequence does it give?

If we start off with 1, 1 as before, the next number will always be 1. But there is another answer. Just as the continuous system had two functions that depended on two initial conditions, this has a second solution too. If we start off with 0, 1, we will get

0, 1, 0.5, 0.75, 0.625, 0.6875, 0.65625 and a lot more decimals.

This will be much clearer if we plot it on a graph.

Exercise 17.1

Run the code at **Series1**. The coefficients should both be 0.5, with starting values 0 and 1.

You should see a set of decaying values alternating about a value of 2/3. The root $k = 1$ means one part of the solution will be a constant value, while the other part will alternate with a factor of 0.5.

$$x = A + B(-0.5)^n$$

and for these initial values we have $A = 2/3$, $B = -2/3$.

Exercise 17.2

Now, with the same starting values of 0, 1, put in coefficients of 1 and −1, then run the code. What do you see? How do you explain it?

We see pairs of values at −1 and pairs of values at +1, with zero values in between them. Perhaps the roots will give the answer.

The equation is now

$$k^2 - k + 1 = 0$$

It now has complex roots, $k = 0.5 \pm j \sqrt{3}/2$

This has a modulus of one, with an *'argument'* that is $+/-\pi/3$.

Sines and cosines are slipping into our solution once again!

Our series is $x_n = \dfrac{2}{\sqrt{3}} \sin(n\pi/3)$.

Exercise 17.3

Run **Series2**. It runs much faster and gives a lot more data points. Enter parameters 1.8 and −0.9, start with values 0, −1 again.

This time you should see a decaying sampled sinewave. To explain it using the roots, we have roots that are those of

$$k^2 - 1.8k + 0.9 = 0$$

$K = 0.9 +/-0.3\,j$, so the roots have a modulus less than one.

We are discovering that for the solution to tend to zero, the modulus of every root must be less than one, but the roots do not have to have negative real parts.

Try parameters −1.8 and −0.9. Now the roots are $k = -0.9 +/-0.3\,j$. The result looks odd, but it still decays to zero. So we can make the following conclusions.

- If a root is complex and its modulus is less than one, it will result in something that looks like a sampled damped sinewave.
- If a real root is negative, again with modulus less than one, the response will switch between positive and negative values on alternate samples, decaying to zero.
- If both roots are positive, but less than one, the response will simply be a decay to zero.

Spend some time exploring coefficient values.

17.3   *Z* AND CODE

Rather than considering subscripts, our interest lies in the way that *z* relates to lines of computer code. If we write

```
x=x+u*dt;
```

we are not writing an equation. We mean that the value of *x* is replaced by its old value, plus *u* times the interval. This is the integrator we have already considered for simulation.

What we really mean is

$$next\ x = x + u * dt$$

or putting *z* into action,

$$zx = x + u * dt$$

$$x = \frac{u * dt}{z - 1}$$

Can we use this technique to analyse our simulations?

We can look more closely at the issues covered in chapter eight, now looking for the implications of the 'next' operator.

Exercise 17.4

Have another look at **Discrete2**. This calculates a crude estimated velocity by subtracting the previous value of *x* from the present one and divides that by the interval.

The code that does that is

```
xtarget=0;
if(t>tnext){
   tnext=tnext+interval;
   vest=(x-xold)/interval;
   u=kx*(xtarget-x)-kv*vest;//Calculate the drive
   xold=x;
}
x=x+v*dt;                   //This is the simulation
v=v+u*dt;
```

So how does *z* help us to analyse this?

xold remembers the present value of *x*, so when the next interval comes around it will be 'previous' *x*. We can denote this by saying that

$$xold = z^{-1}x$$

This code runs with small 'continuous' steps *dt* between the 'interval' steps, so we first need to derive the discrete-time equations. Let us write *T* instead of 'interval'.

For *x* we now have

$$next\ x = x + Tv + \left(T^2/2\right)u$$

while for the velocity we still have

next $v = v + T u$

Ignoring *xtarget*, we can add the 'z's to the simulation assignment statements to get

$(z-1)x = Tv + (T^2/2)u$

$(z-1)v = Tu$

$u = -kx.x - kv(x - z^{-1}x)/T$

We have

$v = Tu/(z-1)$

and putting these together, we get

$(z-1)x = \{T^2/(z-1) + T^2/2\}u$

where

$u = -(kx + kv/T)x + (kv/T)z^{-1}x$

This gives

$(z-1)x + \{T^2/(z-1) + T^2/2\}\{(kx + kv/T)x - (kv/T)z^{-1}x = 0$

Multiplying through by z (z − 1) and sorting out, we find

$\left[(z-1)^2 z + \{T^2 + (z-1)T^2/2\}\{(kx + kv/T)z - kv/T\}\right]x = 0$

$\left[(z-1)^2 z + (z+1)T^2/2\{(kx + kv/T)z - kv/T\}\right]x = 0$

and finally

$\left[z^3 + z^2\{-2 + T^2/2(kx + kv/T)\} + z\{1 + T^2/2 kx\} - ((T^2/2)kv/T)\right]x = 0$

This will give a rather messy cubic polynomial in *k* to solve, so let us substitute the default values from the simulation.

$kx = 4$, $kv = 2$, $T = 0.2$, hence $kv/T = 10$, $(kx + kv/T) = 14$, $T^2/2 = 0.02$

We will get

$k^3 - 1.72k^2 + 1.08k - 0.2 = 0$

An online equation solver found roots

$k = 0.31233$ and $k = 0.70383 \pm j0.38073$

The complex roots have modulus 0.80002, so all three roots have modulus less than one.

In the complex plane they all lie inside the '*unit circle*'.

If we try $kx = 2$ and $kv = 2$, we have

$$k^3 - 1.76k^2 + 1.04k - 0.2 = 0$$

The roots are now $k = 0.41545$ and $k = 0.67228 \pm 0.17162\,j$.

Settling is much faster, but there are still complex roots.

How does this help us to choose feedback values?

In the continuous case, we could improve an oscillatory response by increasing the damping. Though the settling time might increase, stability would not be an issue. For discrete-time control, that is not the case.

The constant term in the equation is $T\,kv/2$. This value will be the products of the roots. So if $kv > 2/T$, there is no way that all their values or moduli can be less than one.

Exercise 17.5

Have another look at **Discrete2**. Try $kv = 10$.

Now set the interval to 0.5 and try to find best feedback values.

Now try interval = 0.05

You should find the task much easier.

17.4   LESSONS LEARNED FROM Z

It is clear that discrete-time control can be a lot trickier than continuous, but when a computer is involved, there is no other option.

The first lesson is the importance of the sampling interval. If good performance is the goal, this should be as short as possible.

Next is the limit imposed by the sampling interval on the magnitude of the feedback. This even applies to our 'pseudo-continuous' simulation with a step length of only a few milliseconds.

In the continuous case, the real part of all the roots must be negative. Increasing the feedback will tend to push them to the left, where they have the whole of the negative half-plane for stability.

In the discrete case, the modulus of all roots must be less than one, meaning that all the roots must lie within the unit circle. Excessive feedback will push one or more roots outside it.

When we apply computer control, we have another obstacle to overcome. That is quantisation.

17.5 QUANTISATION

Although computers can deal with floating-point numbers with vast precision, sensors give readings within a limited range of integers. Even when the quantity is continuous, like the voltage of a tachometer, it must still be converted by an *'analogue to digital converter'* to a digital integer.

Exercise 17.6

In chapter eight you met the simulation **Motor7**. Run it once again. A quantised version of the position, *uq*, represents the position measured to increments of .01. This means that the difference between *uq* and its previous value will also be multiples of .01, which when divided by the time-step $dt = .01$ will give estimated velocity in multiples of one.

As you will have seen, although the estimated velocity is effective for controlling the system, it is of little use if the velocity is to be recorded or displayed.

In the continuous case, we were able to construct a lagged version of the position, and then subtract it from the present value

$$\tau \text{ vest} = \left(1 - \frac{1}{1 + \tau s}\right)x$$

giving

$$\tau \text{ vest} = \frac{\tau s}{1 + \tau s}x$$

We can do the same with the discrete-time system, as you will have seen in **Motor8**.

Exercise 17.7

Run **Motor8** again.

Now you will see a plot of velocity that is very much better, though a little rough. With z transforms in mind, we can apply some analysis.

In the 'loop' window you will see

```
vest=(xq-xqslow)/tlag;
xqslow=xqslow+vest*dt;
```

so we can say

$$z \, xqslow = xqslow + \left(xq - xqslow\right) * dt/tlag$$

i.e.

$$\left(z - 1 + dt/tlag\right)xqslow = xq * dt/tlag$$

and as a transfer function

$$xqslow = \frac{dt/tlag}{z - 1 + dt/tlag}xq$$

thus

$$vest = \frac{z-1}{z-1+dt/tlag} \frac{xq}{tlag}$$

17.6   DISCRETE TRANSFER FUNCTION

When we come to employ vest, we add it to the feedback of (*target-x*). Where before we had just the two feedback gains to select, now we have to select *tlag* as well. Nevertheless, a root-locus diagram can help with the choice, as we will see in the next chapter.

For the simulated motor itself, we have

$$x = \frac{1}{(z-1)^2 dt^2} u$$

where the controlling input is

$$u = kx(target - x) - kv \text{ vest}$$

Now we can substitute for *u*, and then use the transfer function that we have found for it to express vest in terms of *x*. After a struggle with the algebra, we will arrive at a transfer function linking *x* to the input target. I will leave that to the reader.

Instead we will look at a simpler alternative.

In the case where we just subtract the previous value, so that *tlag* = *dt*, this simplifies to

$$u = kx \, target - \left(kx + kv/dt - z^{-1} kv/dt\right)x$$

In the same way that a phase-advance involved subtracting a proportion of a lagged version of the signal, so this involves subtracting part of a delayed version.

Instead of *kx* and *kv*, we can use *a* and *k*, to say

$$u = k\left(target - \left(1 - az^{-1}\right)x\right)$$

To our two poles at $z = 1$ we have added a zero at *a* and another pole at the origin.

This means that as well as subtracting the present *x*, we add a little of its previous value. Will that stabilise the system? The next chapters will tell you.

18 Root Locus

ABSTRACT

Now we add a powerful tool to our armoury, the root locus. We have learned how to express system responses as complex exponentials, determined by roots of an equation that are complex numbers. Now we can see how those roots move as we change the feedback gain.

An additional trick has been added to the conventional root locus plotter. Having chosen a point on the locus that looks like a possible candidate, we can click on it to see the value of the feedback gain that will correspond to it.

A simple feedback filter, such as a phase advance, will usually consist of a pole and a zero. We can move these to find the most hopeful controller,

But the root locus is just as applicable to discrete-time systems, too.

The tilting plank experiment of Chapter 20 challenges us to place two controller poles and two zeros, so a further embellishment enables these to be moved dynamically.

18.1 INTRODUCTION

Let us go back to the continuous system that represents an undamped motor

$$\ddot{x} = u$$

When we put this into transfer function form, using s notation, we have

$$x = \frac{1}{s^2} u$$

In general, the transfer function will be

$$x = G(s)u$$

The question that concerns us is how much negative feedback to apply.

When we apply feedback $u = k$ (target $- x$) to that system, we have

$$x = kG(s)(\text{target} - x)$$

So

$$(1 + kG(s))x = kG(s)u$$

$$x = \frac{kG(s)}{1 + kG(s)} \text{target}$$

DOI: 10.1201/9781003363316-18

To look into stability, we need to know when the denominator becomes zero, i.e. we wish to know the roots of

$$1 + kG(s) = 0$$

This means that

$$G(s) = -1/k$$

For this, $G(s)$ has to be real and negative.

For our undamped motor,

$$\frac{1}{s^2} = -\frac{1}{k}$$

So the roots will be

$$s = \pm j \sqrt{k}$$

There will be no way that we can get stable roots in the left half of the *complex frequency plane* with proportional feedback alone.

18.2   THE COMPLEX FREQUENCY PLANE

The functions of time that correspond to roots in this plane are proportional to e^{st}.

Now s is a complex number, which we can write as $\lambda + j\omega$.

So

$$e^{st} = e^{\lambda t} e^{j\omega t}$$

$$e^{st} = e^{\lambda t} \left(\cos\omega t + j \sin\omega t \right)$$

The real part of this is $e^{\lambda t} \cos \omega t$.

If s is imaginary this is just a sinewave, but if λ is positive, that sinewave will be multiplied by an expanding exponential. It is only if λ is negative that the exponential will be decaying.

Exercise 18.1

Run **Complex1**. This shows a pattern of responses for a few values of s.

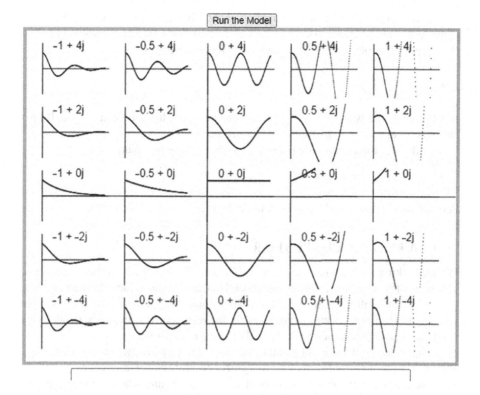

Screenshot of Complex1.xhtml, showing responses for various values of *s*.

For more detail, **Complex2** allows you to click in the complex frequency plane to
see the response for any chosen value.

18.3 POLES AND ZEROES

In looking for the complementary function, we find ourselves looking for the values
of *s* where the denominator $1 + k\,G(s)$ becomes zero. We can plot them as points in
the complex frequency plane and call them 'poles' or 'roots'. As we vary the feedback gain *k*, those roots will move. If we plot their locus, we have more information
to help us choose the best feedback gain to apply.

If we apply zero feedback gain, we are left with the original gain $G(s)$. But this
will have poles of its own. So the roots will start off at those poles, moving away as
we increase *k*.

In the case of our simulation of an undamped motor, we have a transfer function
$1/s^2$, representing two poles at the origin.

When we apply feedback gain *k* to our $1/s^2$ system, we see that the roots are at
$\pm j\sqrt{k}$. They are at symmetric points on the imaginary axis. As *k* varies, the locus
covers the whole axis.

We had some success when we applied phase advance to the feedback. Although in practice we might want to use much shorter time constants, let us consider $(1+3s)/(1+s)$. With this phase advance added to it, $G(s)$ will become

$$G(s) = \frac{1+3s}{(1+s)s^2}$$

In addition to the two poles at the origin, our new gain has a pole at $s = -1$. It also has a 'zero' at $s = -1/3$, where the numerator becomes zero.

To calculate the closed-loop poles we will be looking for solutions of

$$s^3 + s^2 + 3ks + k = 0$$

Perhaps we can harness the power of the computer to do the task for us.

18.4   A ROOT LOCUS PLOTTER

We are looking for the values of s where $G(s) = k$. For this there are two conditions. The first is that $G(s)$ must be real, the second is that the real value must be negative.

It is not difficult to write code that will evaluate $G(s)$, it just needs some simple routines to deal with complex numbers. If we represent two complex numbers as arrays $[a, b]$ and $[c, d]$, their product will be $[ac - bd, ad + bc]$.

So for an array of values of s that cover our complex plane, we can evaluate $G(s)$. We mark each point with a colour or symbol to represent one of four conditions. These are whether the real part is positive or negative, and whether the imaginary part is positive or negative. The root locus will be the boundary between the zones where the imaginary part is positive or negative. Only that section where the real part is negative will count.

Exercise 18.2
Run **Map1**. You will see the s-plane filled with red and yellow. Where the imaginary part of the gain is positive, red 'blobs' have been drawn, with yellow ones otherwise. In some areas, white dots have been drawn at the centres of the blobs. This is where the real part of the gain is positive.

So the boundary between red and yellow shows the root locus, except in the regions with white dots.

We see that with the phase advance we propose, stable control is possible, but the response will be a lightly damped sinewave for any value of gain.

18.5   A BETTER PLOT

The gain map is fine for giving a rough idea of the root locus, but it is not hard to draw a much more precise plot. The explanation that follows will get rather deep in mathematics, so feel free to skip it if you are happy to take it on trust.

The function $G(s)$ defines a 'mapping' from the frequency plane to the gain plane. For each complex value of s, there is a complex value $G(s)$ that gives a point in the gain plane.

What we are trying to do is just the opposite. We are looking for the 'inverse mapping' of the real gain axis into the frequency plane.

We can 'sniff it out' by stepping across the frequency plane looking for the sign of the imaginary part of the gain to change signs. When we have done horizontal sweeps for all the steps of omega, we can sweep vertically in steps of lambda, to fill in any gaps.

The values either side of the change will enable us to interpolate to estimate just where our sweep crossed zero.

Exercise 18.3
Run **Locus1**. You will see spots where the crossing points were interpolated.

But we can do even better! In mapping the real s axis, $G(s)$ rotates it according to the gradient of the function.

We are already evaluating $G(s)$ either side of the crossing, so we can take the difference of those values, divide by its modulus and use that to rotate a horizontal step segment in the opposite direction. We now have an almost continuous curve, except near the origin where gradients are flat.

Exercise 18.4
Run **Locus2** to see the improved version.

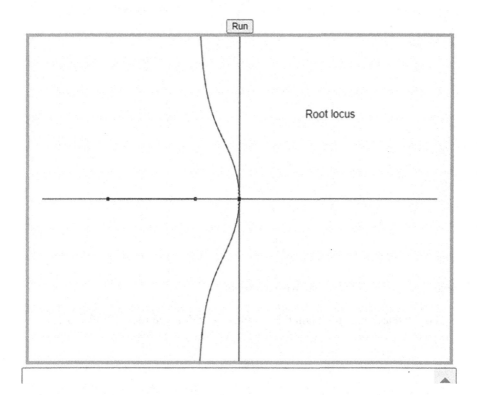

Screenshot of Locus2.xhtml.

But there's more!

We can easily inspect the root locus to choose where we would like the closed-loop roots to lie, but we still do not know what feedback gain we should choose.

The third version allows you to click anywhere in the plane to see the value of $G(s)$ there. If you click on the locus, you should see a value that is real. Your feedback gain will then be the inverse of that value.

Now the code will allow you to enter your own values for the poles and zeroes.

Exercise 18.5
Run **Locus3**. Choose an appropriate value of gain for the best response.

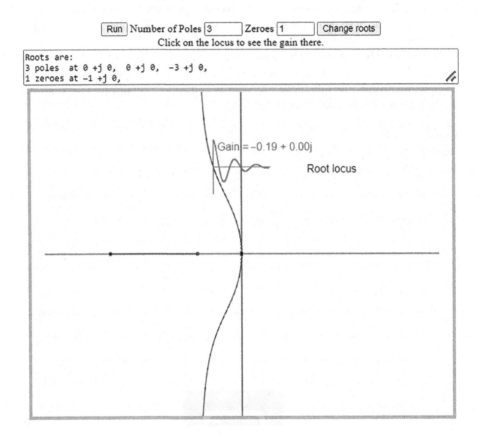

Screenshot of Locus3.xhtml showing a response.

The feedback gain to obtain this response would therefore be $1/0.19 = 5.26$.

18.6   ROOT LOCUS FOR DISCRETE TIME

The root locus technique is just as valid for discrete-time systems as it is for continuous ones. At the end of the last chapter, we found a z-plane transfer function for the control of a system with estimated velocity.

The root locus method will be unchanged except for one test. Rather than roots needing to be in the left-hand half-plane, they now must lie inside the 'unit circle', a circle with radius one, with centre at the origin.

The last chapter ended with:

$$u = k\left(\text{target} - \left(a - z^{-1}\right)x\right)$$

To our two poles at $z = 1$, we have added a zero at $1/a$ and another pole at the origin.

Exercise 18.6
Run **Locus4**. This has been edited from the continuous version and is still full of s notation, but it shows the unit circle.

When you run it, you will see two poles at the origin, one at $(1, 0)$ plus a zero at $(0.85, 0)$.

You should see that the zero has caused the locus to bend round, so that now a good response can be obtained.

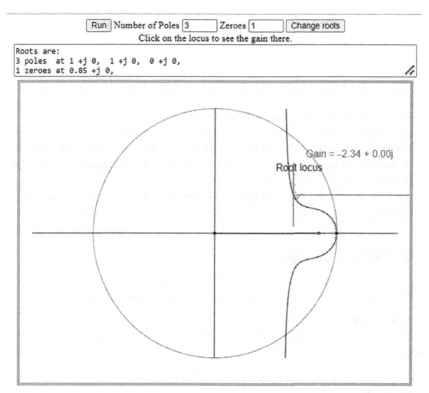

Screenshot of Locus4.xhtml, with a zero at 0.85.

Now change the value of that zero to 0.889 and you will see a strange change. The locus now touches the real axis, so that a non-oscillating response is possible.

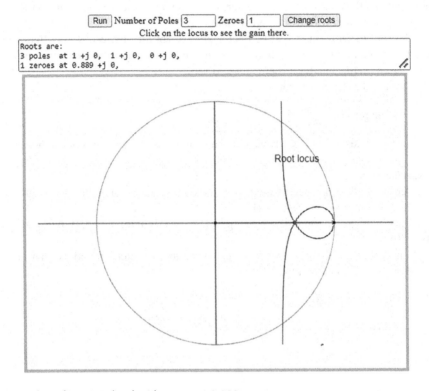

Screenshot of Locus4.xhtml with a zero at 0.889.

A zero at 0.9 would mean that the feedback would consist of x plus nine times the difference between x and its previous value. Noise and quantisation would likely make undesirable any values nearer to one.

Maybe it is worth struggling with the version in the previous chapter that involved *tlag*. We had

$$\text{vest} = \frac{z-1}{z-1+dt/tlag} \frac{x}{tlag}$$

and

$$u = kx\,(\text{target} - x) - kv\,\text{vest}$$

Putting these together we get

$$u = kx\,(\text{target} - x) - \frac{kv(z-1)}{(z-1+dt/tlag)\,tlag}\,x$$

We should define $dt/tlag = b$.

$$u = kx \text{ target} - \frac{(z-1)kv/tlag + (z-1+b)kx}{(z-1+b)} x$$

$$u = kx \text{ target} - \frac{z(kv/tlag + kx) - (kv/tlag + kx(1-b))}{(z-1+b)} x$$

If we define $k = kx + kv/tlag$ we can rewrite this as

$$u = kx \text{ target} - k \frac{z - 1 + b \cdot kx/k}{(z-1+b)} x$$

Stepping back from the task, we see that a pole is being added at $z = 1 - b$, while a zero is added at $z = 1 - cb$. Now c is proportional to the constant ratio that we will maintain between position and *vest* feedback, and c will be less than one.

Let us try putting the pole at $z = 0.55$ and the zero at 0.95.

Exercise 18.7
Run **Locus4** to see the effect of these values.

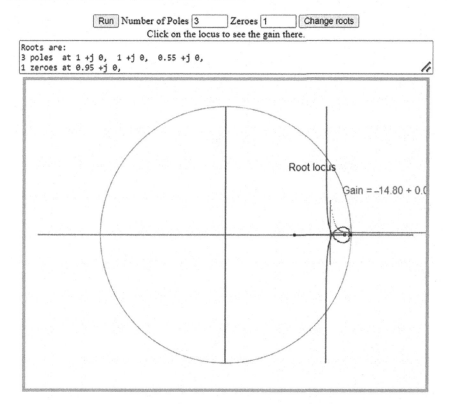

Screenshot of Locus4.xhtml with pole at $z = 0.55$ and the zero at 0.95.

It would appear that a non-oscillating response is still possible, but the simpler estimate that uses the difference between present and previous positions can give a crisper response. Moreover, the point where the locus lies on the real axis has a gain of around fifteen, meaning that the feedback control will be very 'soft'.

18.7   MOVING THE CONTROLLER POLES AND ZEROES

In Chapter 20 you will meet the 'ball-and-beam' experiment, which is third order and discrete time. Two poles and two zeroes are necessary to control it, so another simulation has been written to make it easier to move these.

Exercise 18.8
Run the simulation at **Locus6** to see how the root placement affects the locus. Move pole[0] to −0.75.

You should find that this is a very useful design tool.

19 More about the Phase Plane

ABSTRACT

When pondering a nonlinear feedback scheme, you might find it useful to sketch a set of phase-plane responses by hand, without the use of a computer. When velocity is plotted against position, the state is seen to follow a trajectory of which the slope is given by the system equation. The technique lends itself to a variety of cases, including friction. It also helps to visualise 'sliding mode', where an input switches rapidly to and fro between extremes, to hold the state on a switching line.

19.1 DRAWING PHASE-PLANE TRAJECTORIES

The simulation at **Phase4** is an easy way to obtain a phase-plane plot. But you might want to visualise phase-plane trajectories without the aid of a computer. To draw the trajectory, you need to know its gradient in the phase plane.

Here, we are plotting v against x, in other words the gradient is

$$\text{Slope} = \frac{d\dot{x}}{dx}$$

Some mathematical juggling tells us that

$$\text{Slope} = \frac{\ddot{x}}{\dot{x}}$$

How can this help us? Let us start with an example.

Consider again the system

$$\ddot{x} = -5\dot{x} - 6x$$

When we divide this by $\dot{x}$ we get

$$\text{Slope} = -5 - 6\frac{x}{\dot{x}}$$

Firstly, we notice that on the axis $x = 0$ the slope is -5.

That axis is an *isocline*, a line which all trajectories cut at the same angle.

In fact, any line through the origin will be an isocline, since $\ddot{x}/x$ will have a constant value on it.

DOI: 10.1201/9781003363316-19

So, on the line $\dot{x} = -x$ the slope will be 1, while on $\dot{x} = x$ it will be -11. On the x-axis the slope will be infinite.

So, we can build up a sort of skeleton spider's web of lines through the origin, crossed with ticks to indicate the slope.

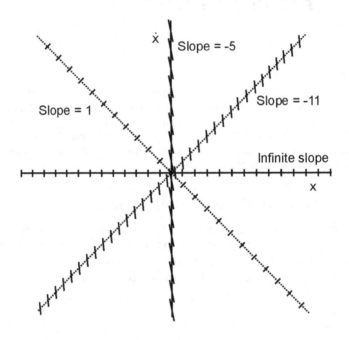

Isoclines.

An interesting isocline is the line $6x + 5\dot{x} = 0$. On this line the acceleration is zero, and so the isocline represents points of zero slope; the trajectories cross the line horizontally.

For this system, there are two isoclines that are worth even closer scrutiny.

Consider the line $\dot{x} = -2x$. Here we find that the slope of trajectories is -2 exactly the same as the slope of the line itself. Once the trajectory encounters this line, it will lock on and never leave it.

The same is true for the line $\dot{x} = -3x$, where the trajectory slope is found to be -3. These are the eigenvectors that we have already met.

Having mapped out the isoclines, we can steer our trajectory around the plane, following the local slope. From a variety of starting points, we can map out sets of trajectories. This is shown in Figure 19.1.

The phase-plane "portrait" gives a good insight into the system's behaviour, without having to make any attempt to solve its equations. We see that for any starting point, the trajectory homes in on one of the special isoclines and settles without any oscillatory behaviour.

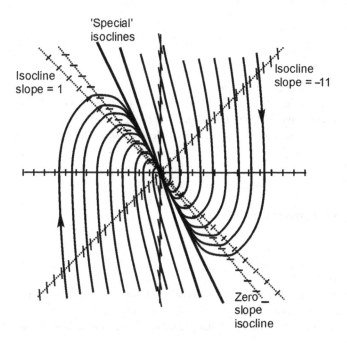

FIGURE 19.1   Example of phase plane with isoclines.

Exercise 19.1
   Sketch the phase plane for the system

   $$\ddot{x} + \dot{x} + 6x = 0$$

The velocity feedback has been reduced so that there are no real roots. The trajectories will match the "spider's web" image better than before, spiralling in to the origin to represent system responses which now are lightly damped sine waves.
Check your result by using the simulation **Phase4**.

19.2   PHASE PLANE FOR SATURATING DRIVE

We are still considering a simple system where acceleration is proportional to the input.

   $$\ddot{x} = u.$$

Assuming zero demand, we are feeding back a mixture of position and velocity

   $$u = -6x - 5\dot{x}$$

On the lines

   $$-6x - 5\dot{x} = +1$$

and

$$-6x - 5\dot{x} = -1$$

the drive will reach its saturation value.

Between these lines the phase-plane plot will be exactly as it was in Figure 19.1. So, we can select just that section, as in Figure 19.2.

What happens outside that region is a different story. The input is now a constant value of +1 or −1.

$$\ddot{x} = -1$$

in the region to the right, and

$$\ddot{x} = 1$$

in the region to the left.

The slope of the trajectories will now be $1/\dot{x}$ or $-1/\dot{x}$. In both cases the isoclines will be lines parallel to the x-axis, on which $\dot{x}$ is constant.

The trajectories will be parabolae that have a horizontal axis. By integrating the equation

$$\ddot{x} = 1$$

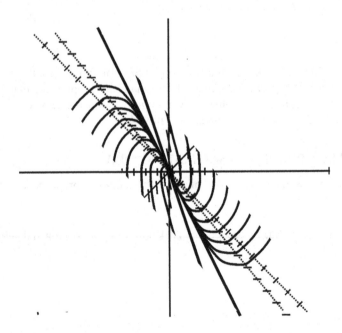

FIGURE 19.2   Linear region of the phase plane.

twice with respect to time, we see that

$$x = \dot{x}^2/2 + \text{constant}$$

with a horizontal axis which is the x-axis.

Similarly, the trajectories to the right, where $u = -1$, are of the form

$$x = -\dot{x}^2/2 + c$$

The three regions of the phase plane can now be cut and pasted together to give the full picture of Figure 19.3.

This figure was generated from **Phase5**. In the shaded region, the trajectories are parabolae and the isoclines are horizontal lines. The line in the centre of the linear zone is an isocline on which the slope is zero.

Exercise 19.2

Run **Phase5** with various values of the damping term. Click to start new trajectories.

We will see that the phase plane can become a powerful tool for the design of high-performance position control systems. The motor drive might be proportional to the position error for small errors, but to achieve accuracy the drive must reach its limit for a small displacement. The "proportional band" is small, and for any substantial disturbance the drive will spend much of its time saturated. The ability to design feedback on a nonlinear basis is then of great importance.

Exercise 19.3

The position control system is described by the same equations as before,

$$\ddot{x} = -5\dot{x} + u$$

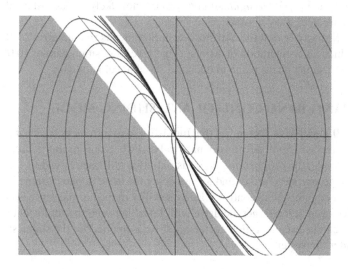

FIGURE 19.3 Phase plane showing linear and shaded saturated regions.

where

$$u = -6x$$

The damping term $5\dot{x}$ is not given by feedback, but by passive damping in the motor. In the unsaturated region, the equation is still

$$\ddot{x} + 5\dot{x} + 6x = 0$$

Once again the drive u saturates at values $+1$ or -1, but this time the boundaries are $x = 1/6$ and $x = -1/6$.

Sketch the new phase plane, noting that the saturated trajectories will no longer be parabolae.

Exercise 19.4

Consider the following design problem, which was previously set in chapter six.

A manufacturer of pick-and-place machines requires a linear robot axis. It must move a one kilogram load a distance of one metre, bringing it to rest within one second. It must hold the load at the target position with sufficient 'stiffness' to resist a disturbing force, so that for a deflection of one millimetre the motor will exert its maximum restoring force.

Steps in the design are as follows:

1. Calculate the acceleration required to perform the one metre movement in one second.
2. Multiply this by an 'engineering margin' of 2.5 when choosing the motor.
3. Determine feedback values for the position and velocity signals.

You might first try *pole assignment*, but you are not likely to guess the values of the poles that are necessary in practice.

You will discover that the 'stiffness' criterion means that an error of 10^{-3} metres must produce an acceleration of $10\,\text{m/s}^2$. That gives a position gain of 10,000.

You can use the phase plane to deduce a suitable velocity gain.

19.3   BANG-BANG CONTROL AND SLIDING MODE

Exercise 19.2 will have shown you that the stiffness requirement means that the drive is mostly saturated, with only a very narrow band where it is proportional. In the case of that example, the *proportional band* is only two millimetres wide.

We can go one step further and always apply the drive at one extreme or the other. This is termed *bang-bang* control. Clearly stability in the normal sense is now impossible, since even at the target the drive will 'buzz' between the limits, but the quality of control can be excellent. Full correcting drive is applied for the slightest error.

Consider the system

$$\ddot{x} = u$$

where the magnitude of u is constrained to be no greater than 1.

If we apply a control law in the form of a logical switch,

```
if ((x+xdot)<0){u=1;}else{u=-1;}
```

then there will be the same two sets of parabolic trajectories as before. The switching line that will now divide the full positive and negative drive regions is the line

$$x + \dot{x} = 0$$

Now just below the point (−.5, .5) we will have $u = 1$. The slope of the trajectory will be $\ddot{x}/\dot{x}$ which at this point has value 2. This will take the trajectory upwards across the switching line into the negative drive region, where the slope will be −2. This will take the trajectory back across the switching line yet again. The drive buzzes to and fro, holding the state on the switching line. This is termed *sliding*.

But the crossing point does not remain fixed. On the switching line, we have

$$x + \dot{x} = 0$$

This is an equation that describes an x that decays as Ae^{-t}. Once sliding has started, the second-order system will now behave like a first-order system. This principle is at the heart of *variable structure* control.

19.4   MORE USES OF THE PHASE PLANE

Fashions of control strategy come and go. At one time a variety of fuzzy control was proposed in which feedback variables were quantised into classes such as 'large, small, near zero, small negative, large negative'. This would divide the phase plane into 25 combinations of velocity and position. A different value of drive could then be assigned to fill each of these regions.

In some systems there is a discontinuity, such as friction where the force reverses with the velocity. The phase plane is then an obvious choice for predicting the behaviour of friction damping. But simulation is likely to be a simpler solution.

20 Optimisation and an Experiment

ABSTRACT

If a rapid response is the objective, 'bang-bang' control is the answer. The input takes extreme values, switching each way until the target is reached. Other forms of optimisation such as the 'two endpoint problem' would involve the intricacies of the 'calculus of variations', but there is none of that here. Instead, a simple technique termed 'Logical Predictive Control (LPC)' is presented, which uses two fast models in parallel. We see that near-time-optimal control can readily be achieved for systems up to fifth order, and maybe beyond.

Research on predictive control began in Cambridge over half a century ago, being demonstrated by creating an experiment involving a ball running in a groove in a tilting plank. This is now widely known as 'ball and beam'. A variation is suggested where a video camera is used to detect the ball position. The point is that we can then attack the solution with our tools of discrete-time analysis, including the root locus. The tools of chapter 18 come into force.

The reader can waste a lot of time, playing with the model here.

20.1 INTRODUCTION

With techniques for ensuring stability well-established, research in control turned to finding the 'best' way to do it. At the heart of optimal control is the concept of a *cost function*. For a mechatronic system this can simply be time. The best pick-and-place robot could be the one that reaches the target fastest.

But other cost functions have long been fashionable, such as the integral of squares of variables. These relate more to the *regulator* problem than to motion, where some industrial process must be controlled to resist disturbances. By using a quadratic function as the cost, the LQR linear quadratic regulator finds an optimum where the controller is linear.

In effect the goalpost has been moved to a location where the ball is already heading.

A major problem is defining that cost function in a way that will best please the client. When an autopilot levels the turning aircraft from a bank angle, I am told that pilots prefer a 'single overshoot' response so that it is clearer to them when the turn is complete.

Another fruitful area for research in optimal control is the 'two endpoint problem'. This has obvious application in weaponry, where you want a pursuit missile to explode as close to its moving target as possible. But with autonomy taking hold of cars and agricultural machinery, new horizons are opening. The self-parking car could be a classic example.

DOI: 10.1201/9781003363316-20

20.2   TIME-OPTIMAL CONTROL

Over half a century ago, L.S. Pontryagin pointed out that to move in minimum time you must always use maximum drive. The tricky task for the controller is to switch from one extreme to the other at the right time.

His 'maximum principle' said that you need one less switch than the order of the system, so a five-cascaded-integrator system would require just four reversals of drive. But the real problem is that a strategy that optimises time is seldom the best one in practice.

For soft-landing a capsule on the moon, a strategy that is both time and fuel optimal is to leave switching on the retro jets until the last possible moment, then to fire them at full thrust until you come to a halt at the surface. But if the jets are a second late igniting, or if your thrust calculation is the slightest bit overoptimistic, the landing will be far from soft.

The time to start firing is calculated from the maximum thrust available. If a 'pessimistic' value is entered into the calculation, firing will start early. But as velocity starts to fall more rapidly than anticipated, thrust can be reduced to bring it back onto the calculated path. A safe landing is much more likely than if full thrust is required for the whole descent.

In the predictive strategies that follow, such 'slugging' can be applied when resetting the states of the model to those of the plant.

20.3   PREDICTIVE CONTROL

In the 1960s, research in Cambridge investigated the use of a fast model for applying optimal control. At the simplest second-order level, the model calculated where the position would come to rest if brought to a halt. If this was short of the target, acceleration would be applied instead of braking.

An analogy might be a tearaway driver trying to get to a red traffic light in minimum time. With foot down hard on the accelerator, he estimates where he would come to rest if he slammed on the brakes right now. If that estimate is short of the halt line, he keeps on accelerating, otherwise he screeches to a halt.

My own research on the three-integrator case was complicated by the need to add a switch to the model drive, which must be calculated to bring velocity and acceleration to zero at the same time. But by considering suboptimal control, I found was a strategy that avoided that switch.

More recently, stimulated by the ease of digital simulation, my team has developed a strategy termed 'Logical Predictive Control (LPC)', which has shown itself capable of controlling systems of cascaded integrators of order at least five.

An example of a multi-integrator system is an aircraft trying to land on the centre of a runway. The input is the aileron, but first we must consider the motor that drives it.

- The integral of the aileron motor velocity is the aileron angle.
- This applies a couple to the aircraft to roll about its length, so its integral is the roll rate.
- The integral of the roll rate is the angle at which the plane is banking. That causes the path of the aircraft to turn.

- So the integral of the bank angle is the heading.
- The integral of the heading is the rate at which the plane is moving across the runway.

And finally, it is that error from the centre of the runway, with all the other variables, that must be fed back to the aileron motor to do the steering!

I hope you feel comfortable on your next flight.

The operation of LPC is as follows:

From an initial state that corresponds to the present state of the system, two simulations are run simultaneously with either extreme sense of the drive. In either case, there will come a time when all the model variables have the same sign as the input drive. Until then, that simulation will be termed 'offside'.

When the model for one or other signs of input ceases to be 'offside', then it is the other sense of drive that is applied to the system. The strategy is then to apply the opposite sense of drive to the plant. It is that simple.

In effect we are applying the drive that causes the model state to remain offside until furthest ahead in time, but when one of the models comes 'onside' there is no point in going further.

Exercise 20.1
If this seems to be too simple to be true, run the simulations that you will find at **Predictive1** to see it in action.

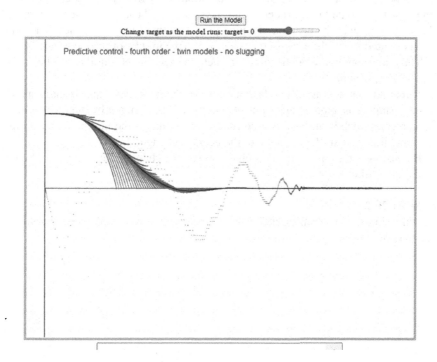

Screenshot of Predictive4twin.xhtml for a fourth-order system.

20.4 A TILTING PLANK EXPERIMENT – NOSTALGIA

To test the effectiveness of the strategy, an experiment was constructed. This consisted of a plank of wood, pivoted in the middle, with a v-notch cut into its length in which ran a large, heavy ball bearing. A motor tilted the plank by means of a rack and pinion, so this adequately demonstrated a third-order system. The three state variables were the plank angle, ball velocity and ball position. Position was measured by stretching a resistance wire along one wall of the V. A copper wire was stretched along the opposite wall, to pick off the potentiometer voltage when the wires were bridged by the ball.

Control was applied with the aid of an analogue computer, constructed from transistor operational amplifiers of my own design.

The machine was exhibited in the Festival Hall in 1966, as part of the celebration of the Third IFAC conference. In 1968, it even found its way into the 'Cybernetic Serendipity' exhibition at the London Institute of Contemporary Art! It was given the name CUPID, 'Cambridge University Predictive Iterative Device'.

Since then that type of experiment has been described as 'Ball and Beam' and has been used widely around the world, but there is a variation that is of interest here.

20.5 BALL AND BEAM: A MODERN VERSION

An easy version of the experiment for any mechatronics lab is to use a camera to detect the ball position, rather than the potentiometer wire. The track can now be a simple strip of folded or extruded metal. It is painted black. A webcam is mounted above it, to observe the position of a ping-pong ball that runs in the track. The camera is oriented to put the view of the track across the middle of the landscape frame.

There are now many publications on such an experiment, but the aim here is to explore simple approaches.

Although I am not usually in favour of stepper motors, this experiment is one for which a stepper is appropriate. Control actions will be dictated by the frame rate of the camera, which is an ideal way to time the stepping of the motor. With a simple lever and linking bar, the tilt angle of the beam is changed.

But just as in the case of the two-phase sensor, the change is incremental. There is no datum until one is established.

Exercise 20.2
Visit the simulation at **Ballbeam1**. Use it to help you to write code for controlling a real version.

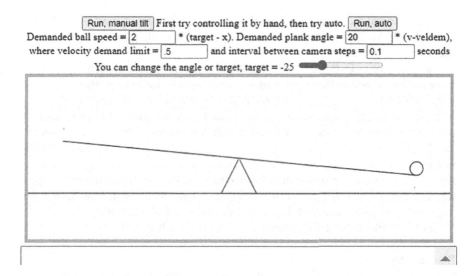

Screen grab of Ballbeam1.xhtml.

Exercise 20.3

When you have adjusted the parameters to obtain the best response you can, visit the simulation at **BallbeamPred** to see what LPC can achieve. Try it at various sampling intervals.

Of course, those version are bogus in practice! We do not know the datum angle of the plank, and the velocity can only be deduced from the difference between successive values of x. The only input to the motor is whether to step or not, and in which direction. Is the situation hopeless?

Fortunately, we can bring the guns of discrete time to bear. In **Ballbeam2** we can measure x and remember its two previous values. Let us call the values x_0, x_1 and x_2. The code presents you with three boxes into which to enter the gains, and then the motor steps according to the sign of their sum, *whichway*. The orange dots show the value of *whichway*, to help you make improvements to the feedback parameters. Control is complicated by the nonlinear nature of the drive limitation.

Exercise 20.4

See if you can find parameters to optimise the **Ballbeam2** simulation. Note that the numbers will be large, since to represent velocity, differences must be divided by the interval.

Are you having difficulties? Can the root-locus plotter at **Locus4** help? The system will represent three poles at $z = 1$, while this controller will contribute two more at $z = 0$. The mix of feedback coefficients will determine three zeroes wherever you care to locate them.

But there is a more pragmatic approach. Our difficulties seem to come from the lack of a feedback term involving the angle. Although we cannot measure it, we can estimate the angle by summing the step changes applied to the motor. Of course, the discrepancy between the estimate and the real angle will cause the ball to come to rest at the wrong position. (For the initial tilt here, it is off the end of the track.) But if we apply a 'decay' to that estimate, the resting place of the ball will decay towards the target.

The parameters in **Ballbeam3** have been rather arbitrarily chosen, but they do bring the ball to rest at the target. I am sure that you can improve on them.

So why does this succeed, while the earlier approach seemed impossible?

Note that we are not even using x_2, the value before last. Instead of two controller poles at $z = 0$, we have one at $z = 0$ and the other determined by the angle estimate lag. Some more investigation with the root-locus plotter can perhaps suggest some improved values.

Another complication is that perhaps we might be looking for a part of the root locus where the gain is positive, rather than negative. So, the locus plotter has been modified so that the positive part of the locus is shown in red, see **Locus5**.

But the black plot with the default values looks most hopeful!

Exercise 20.5

How can those values be translated into feedback coefficients? I leave that one to the reader.

A root locus is a useful tool when a single variable is to be explored, the feedback gain. But now we can have two poles and two zeroes to manipulate while we explore possible controllers. The simulation at **Locus6** addresses this by providing four slide bars to change them, rather than fiddling with numbers. As each value is changed the plot appears instantly.

Exercise 20.6

Run **Locus6** and drag the control bars to see the plot change. Can you find better feedback values?

The conclusion is that when the system is nonlinear, perhaps because of a limited drive, a nonlinear controller can give much better results. The predictive strategy is seen to correct an error in close to minimum time.

The ball-and-beam experiment and its predictive controller were given the name CUPID, Cambridge University Predictive Iterative Device, when exhibited in the London Festival Hall for the 1966 IFAC conference. A set of posters, Figure 20.1, were designed by my wife, Rosalind, to explain its operation.

Another poster summed it all up.

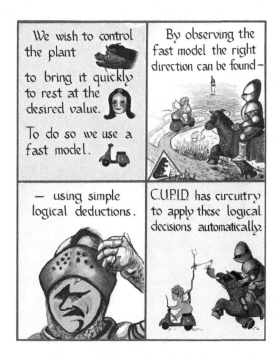

FIGURE 20.1 Posters illustrating C.U.P.I.D. for IFAC 1966.

A poster summing up LPC.

21 Problem Systems

ABSTRACT

We have looked at systems where control is applied at discrete intervals, but systems containing a time delay present their own special problems. A laboratory experiment is described where water must travel through a metre of tubing before its temperature is measured. Those taking a shower where there is a long tube from the mixer will appreciate the problem when a kitchen tap is suddenly turned on, sending a shiver down their spine.

Several simulations show the virtue of integral action, restoring the output to its exact desired setting, but the response might seem slow to our shivering bather.

21.1 INTRODUCTION

To reduce the final settling error, it is natural to wish to increase the feedback gain. But there are systems where stability strictly limits the gain that can be applied. The answer can in many cases be the use of *integral action*. Instead of feeding back the error immediately, it is integrated to be fed back slowly. Though a disturbance might take time to be corrected, the integral will eventually ensure that it is reduced to zero.

The integral term adds the middle letter to the traditional technique of 'PID' (Proportional plus Integral plus Derivative control). Tuning the right mixture has long been the task of the control technician.

21.2 A SYSTEM WITH A TIME DELAY

Any time delay can add a special problem to control. Even the short delay of a flexible pipe leading to a wall-mounted shower can give the experience of instability when manually trying to set the right temperature.

Figure 21.1 illustrates a simple control experiment which I devised around 1970. An electric kettle element heats water in a container, the water then flows through a metre of tube before its temperature is measured.

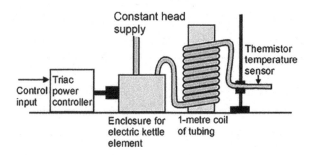

FIGURE 21.1 Water heater experiment.

DOI: 10.1201/9781003363316-21

When we try to find a transfer function for it, we run into difficulties. The enclosure poses no great problems. Water flows into it at ambient temperature and out at the tank temperature. So, the tank temperature is a simple first-order lag. Its inputs are the heating power and the temperature of the incoming water. The tank temperature can be simulated with:

```
tank = tank + (a * (ambient - tank) + b * u) * dt
```

where parameter *a* depends on the flow rate and parameter *b* depends on the heater power.

It is in the pipe that the main problem arises. In effect, the temperature of every specific drop of the water in transit can be regarded as a state variable. In the simulation that follows, this has been approximated by one hundred sections. This is easy to simulate, but a transfer function is out of the question.

The simulation at **Heater1** updates the variables at 0.1 second intervals. With a hundred elements in the array, the pipe delay amounts to ten seconds. The time constant of the tank has also been set to be about ten seconds.

Exercise 21.1

Run the simulation and verify that the response is as shown in Figure 21.2. The ambient temperature will also change as you run your mouse across the bar.

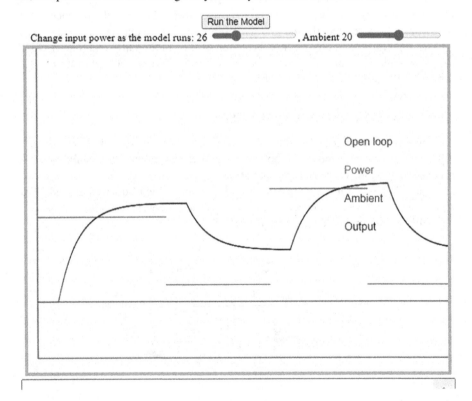

FIGURE 21.2　Response of the simulation. Note the delay.

If the calibration of the heating effect were known accurately, the temperature could be controlled by *feed-forward* alone, merely setting the heating to a proportion of the required temperature rise and waiting for the output to respond. But that relationship will depend on the flow rate, and there is also any change in the ambient temperature to contend with.

Exercise 21.2
Run the simulation at **Heater2**. Adjust the gains to get the best response you can. Note that for stability the feedback gain is strictly limited, so that ambient variation is largely uncorrected.

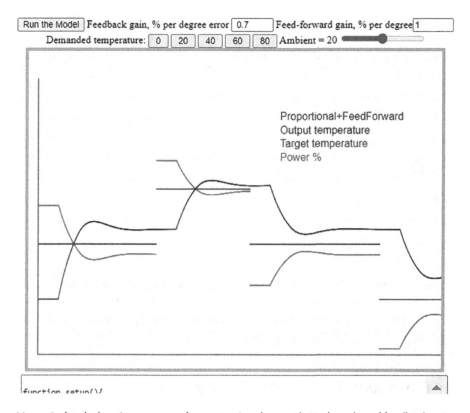

Heater2.xhtml, showing response for proportional control. Find a value of feedback gain for which the response is reasonably well damped.

21.3   INTEGRAL ACTION

Now we introduce integral action. The temperature error is integrated by adding a proportion of it at each step to the variable *integral*. The input is now the sum of the error and its integral. The proportional gain and the time taken for the integration must be carefully balanced to obtain a satisfactory response.

Exercise 21.3
Run the simulation at **Heater3**.

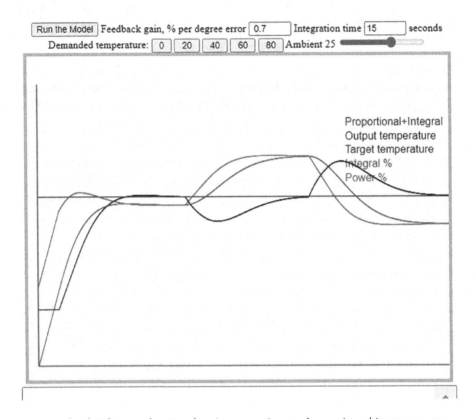

Heater3.xhtml with integral action, showing corrections to changes in ambient temperature.

Note that it is capable of correcting for changes in the input water ambient temperature within about thirty seconds. Can you find values that will make the response any faster?

21.4   THE BATHROOM SHOWER APPROACH

In a bathroom shower, if the pipe from the mixer is long, there will be a delay before a tap setting reaches the bather. It is a good idea to wait a couple of seconds to let the temperature settle before changing the setting. The simulation at **Heater4** shows this in action.

22 Final Comments

ABSTRACT

As the book draws to a close, it is clear that control theory covers a much wider span than the mechatronic student will usually need. Hopefully, the essentials covered here will address the majority of problems to be encountered, or at least give a good starting point.

A tip is finally included to enable a microcontroller to handle several two-phase encoder signals at the same time. When a small DC motor is running at full speed, with a high-resolution encoder, updates must be made at very brief intervals. Some assembler-level instructions make it possible.

22.1 INTRODUCTION

There are many powerful mathematical techniques for the analysis and design of control systems. Some are based in the frequency domain, with heavy use of complex numbers. Some time-domain methods lean heavily on the algebra of transformations. But in many practical systems, the design requirements are dominated by physical aspects such as drive limitation and backlash. The final test must be the construction and examination of a prototype system, although simulation can go a long way to achieving a result.

Computers have evolved from the concept of a central 'thinking machine' to a network of collaborating processors. Though simple mechatronic devices can rely on a single processor, motor cars now abound with them. An important task is then their communication structure, in which devices might operate at a variety of clock rates.

As soon as a computer is introduced into a control loop, discrete time effects must be considered. But a more significant impact of discrete time arises when video is used as the sensor. The intervals between control decisions cannot be less than the value dictated by the frame rate.

The mathematics of discrete time control are mostly based on action at constant intervals, but control can take many forms. A multitude of systems apply 'control when needed', without a regulating clock rate. The most common example of this is the thermostat. This is one of many cases where simulation can outperform algebra.

22.2 MULTI-RATE SYSTEMS

The 'velodyne loop' has already been mentioned as a way that velocity control can be applied with a high gain, to 'stiffen' a control loop. The first-order loop presents few stability problems. A signal is then injected into the velodyne loop as a demand for velocity. If this demand is generated at discrete intervals, they do not have to be as frequent as any steps used for the inner loop.

A recent project concerned the control of a tower crane. It used a camera to observe the swinging of the load. For the laboratory model, time constants were very

DOI: 10.1201/9781003363316-22

much shorter than would be found on a 'real' crane, while the camera frame rate was still limited to 25 fps. For reasons that will be described, the cycle time of the motor control loop had to be very short indeed. The combination of this with an outer loop of 40 milliseconds proved to be successful.

Here is a method for controlling a motor with an encoder, that you may well find useful.

22.3   MOTOR CONTROL WITH A TWO-PHASE ENCODER

The model was moved using DC motors. They carried two-phase incremental encoders, as described in chapter nine.

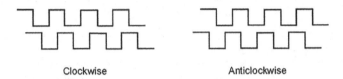

Clockwise   Anticlockwise

Two-phase encoder signals.

These made it possible to measure displacement with great precision, but presented problems of their own. The encoders gave 48 transitions per rotation of the motor. With a top speed just under 200 revolutions per second, this meant that the loop to read the encoder must be no longer than 100 microseconds.

The software of the Arduino microcontroller is optimised for hobbyists. It is based closely on the language 'C'. The code to input a logic state is expressed in terms of reading each individual pin. In fact, at machine level an 8-bit byte is input. Time-wasting code is used to extract that single bit from it.

Fortunately, Atmel assembler commands can be included, too. With a line such as

```
bits=PORTC;
```

a whole byte is read in one instruction. If we use that to input signal pairs from three motors, our task is then to interpret this byte to control each motor.

For clockwise rotation, the bit pairs of each motor cycle in the order $10 > 11 > 01 > 00$. In the strategy that follows, the previous pair are shifted two places left and 'or' ed with the new values. This results in four 4-bit 'nibbles': 1011, 1101, 0100 and 0010 will represent a clockwise movement and 1101, 0111, 0001 and 0010 will mean an anticlockwise change. The combinations 0011, 0110, 1001 and 1100 will signify an error, meaning that the routine has been so slow that both bits have changed. The remaining 0000, 0101, 1111 and 1010 will mean that there has been no change.

A simple 16-element array can give the value of the change in position that will be added to the current position of each motor.

```
int move[16] = {0, 1, -1, 0, -1, 0, 0, 1, 1, 0, 0, -1, 0,
-1, 1, 0};
```

For the first motor, with position stored in *m1pos*, we only need

```
change = ((oldbits & 0b1100) | (bits & 0b11));
m1mov = move[change];
m1pos += m1mov;
```

For the second motor, a different pair of bits are selected and a shift is added

```
change = ((oldbits & 0b110000) | (bits & 0b1100))>>2;
```

while for the third motor

```
change = ((oldbits & 0b11000000) | (bits & 0b110000))>>4;
```

Finally, we shift and remember the bits for the next test

```
oldbits=bits<<2;
```

To speed the process even more, it is called as an interrupt routine. This is set up by

```
cli();
PCICR |= 0b00000010; //turn on port C interrupts
PCMSK1 |= 0b00111111; //enable PC0- PC5, Arduino pins A0 to A5
sei();
```

That same Arduino was entrusted with the task of controlling the drives to the motors, to achieve the commanded velocities.

22.4   AND FINALLY

Devices such as the Arduino and the BBC Micro:bit are simple enough that software can be traced down to the fundamental assembly instructions. The programmer can expect code to be executed in a predictable way.

As the operating systems of personal computers and mobile phones become more sophisticated, any application must compete with rival processes that often number in the hundreds. Internet connectivity means that there must be substantial resources devoted to the detection and suppression of malware. There will also be frustrating limits and blocks on the tasks that user-generated software is allowed to perform.

As the thrust of research in control theory moves further into self-correction and 'deep learning', troubleshooting the software might soon resemble psychoanalysis.

Now Read On

Rather than put references at the end of each chapter, they are gathered here so that some review remarks can be added.

CHAPTER 1

Although principles of control theory had been used down the ages, such as windmills that rotated to face the wind, it was in the Second World War that it started to be seen as an academic subject in its own right.

I am normally derisive of Wikipedia references, but a good description of the ball-and-plate integrator as used in a bomb aimer can be found at:

Wikipedia. (2022). Ball-and-disk integrator, viewed January 2023 at https://en.wikipedia.org/wiki/Ball-and-disk_integrator

For a long time, electrical engineers had been working to stabilise amplifiers using frequency domain methods, applying sine waves to a system and observing the amplitude and phase of the output. Soon their efforts were absorbed into control theory.

I summarised some of their techniques in chapter 11 of:

Billingsley, J. (2010). *Essentials of control techniques and theory.* Boca Raton, FL: Taylor & Francis.

Those methods are unlikely to be of use to mechatronics experimenters, but the root locus techniques of its chapter 12 have been greatly enhanced in this book.

CHAPTER 2

Now we look at dynamic systems from the 'state-space' point of view, where instead of manipulating transfer functions we pay attention to the 'state variables' that describe what is happening.

Analogue computers had already given clues to their importance, but the theory of state variables was drawn together by:

Zadeh, L. A. & Desoer, Charles A. (1963). *Linear system theory: The state space approach.* New York: McGraw-Hill.

This was my essential reading as I prepared to return from autopilot design to Cambridge research. It was a rather intellectual approach, but provoked a spate of books and papers with the theme of 'Modern Control Theory'.

Quick on the scene was Julius Tou:

Tou, J. (1964). *Modern control theory.* New York: McGraw-Hill.

followed not long after by K. Ogata:

Ogata, K. (1967). *State space analysis of control systems.* Englewood Cliffs, NJ: Prentice Hall.

These were long regarded as classics, leading to some further editions and updates:

Ogata, K. (2000). *Modern control engineering.* Englewood Cliffs, NJ: Prentice Hall, ISBN, 13, 978-0.

Many authors followed in their footsteps, drawing material together for the many new courses that were appearing in control theory. My effort a little later was:

Billingsley, J. (1989). *Controlling with computers: Control theory and practical digital systems.* New York: McGraw-Hill College, ISBN 13: 9780070841932.

CHAPTER 3

The main reference here is to the website www.jollies.com, set up decades ago to introduce JavaScript simulation. It greatly needs tidying up.

JavaScript keeps on being updated, largely because of security concerns, but there is a lot of information at:

https://developer.mozilla.org/en-US/docs/Web/JavaScript/Guide

CHAPTER 6

I do not think that chapters four and five need a reference.

At some time in the last century, I had a consultancy with IBM. They were having problems with setting feedback values for a precision plotter, that was persistently overshooting. As a result, I realised that drive limitations were more important than poles and zeros. I recall punching out a program as a stack of punched cards and taking it to IBM Hursley to run – with just one error. The solution worked.

It resulted in a paper:

Billingsley, J. (1991). On the design of position control systems, *IEE Proceedings D (Control Theory and Applications)*, Volume 138, Issue 4, pp. 331–336.

CHAPTER 7

The bicycle problem has much in common with that of the autopilot. Many of the pragmatic techniques of introducing limits into the controller arose from that early industry experience, rather than doctoral research.

A web search for 'nonlinear autopilot' will bring many hits.

CHAPTER 9

This chapter moves away from theory and considers sensors and actuators.

In 1983 I published a book for hobbyists,

Billingsley, J. (1983). *DIY robotics and sensors with the BBC computer*. London: Sunshine Publications.

It was a huge success, spawning another version:

Billingsley, J. (1984). *DIY robotics and sensors on the commodore computer: Practical projects for control application*. London: Sunshine Publications.

This was also translated into German and Spanish.

Please forgive me for citing a video review that I found on the web:

https://www.youtube.com/watch?v=Bh_ydIA3sIo

But these did not really have much academic merit, so later I published something more formal:

Billingsley, J. (2013). *Essentials of mechatronics*. Hoboken, NJ: John Wiley, ISBN: 978-1-118-73803-0.

But you will find many more textbooks on the subject with a web search.

CHAPTER 10

To find out more about operational amplifiers, a web search pulls up a host of helpful articles – many by companies selling their product.

But, of course, there is a huge amount in books, such as:

Bird, J. (2013). *Electrical circuit theory and technology*, 5th ed. New York: Routledge. https://doi.org/10.4324/9781315883342

CHAPTER 11

This goes more deeply into the task or manipulating and unravelling state equations, so it might be best to refer you back to the references for chapter two.

CHAPTER 12

Now we return to the task of controlling things, recalling tasks like the inverted pendulum. You will find a whole chapter on this in:

Billingsley, J. (2010). *Essentials of control techniques and theory*, 1st ed. Boca Raton, FL: CRC Press. https://doi.org/10.1201/b15297

already cited. But many others have presented convoluted solutions, some of which may not be very helpful.

You will have seen that the idea of 'pole assignment' is a simple one. You fiddle your feedback gains to put the roots where you think you need them. Now the method is embodied in packages such as MATLAB. You can see a useful description at:

https://www.mathworks.com/help/control/getstart/pole-placement.html

CHAPTER 13

Until around 1960, the standard textbook for differential equations was:

Piaggio, H. T. H. (1949). *An elementary treatise on differential equations*. London: G. Bell and Sons Limited.

You can still read it at:

https://ia800702.us.archive.org/16/items/elementarytreati032501mbp/elementarytreati032501mbp.pdf

Heaviside notation had been used until then for dealing with differential equations, where the operator d/dt was simply replaced by the symbol D. The function '1' was a unit step at $t = 0$.

Then the Laplace transform became fashionable. I could see few advantages and numerous disadvantages. Admittedly you could treat 's' as a simple variable rather than an operator. But the function '1' was now an infinite spike. An extra 's' crept into all equations.

In both cases, equations were not really solved. Instead, expressions were matched with the transforms of known functions, "Here's one that I transformed earlier." But very rapidly, textbooks on control theory were filled with "Everything you need to know about the Laplace transform."

CHAPTERS 16, 17 AND 18

All three chapters are illuminated by the theory of complex variables. My grounding in complex variables came from:

Copson, E. T. (1935). *An introduction to the theory of functions of a complex variable*. Oxford, UK: Oxford University Press.

It can still be read at

https://ia601404.us.archive.org/24/items/in.ernet.dli.2015.233836/2015.
233836.Theory-Of.pdf

I cannot remember my other textbooks for the Mathematics Tripos.

CHAPTER 20

When I returned to academe in 1964 with my new bride, L. S Pontryagin had set the control world alight with his 1956 Maximum Principle:

Pontryagin, L. S., Boltyanskii, V. G., Gamkrelidze, R. V., & Mishchenko, E. F. (1963). *The mathematical theory of optimal processes*, K. N. Tririgoff, Transl., L. W. Neustadt, Ed., New York: Wiley.

Its essence was that to achieve a target in minimum time, the input must be at one extreme or the other. It still left the problem of knowing when to switch. Harold Chestnut, another IFAC founder, had considered a predictive strategy:

Chestnut, H. & Wetmore, V. (1959). 'Predictive control applied to a simple position control'. Engineering Laboratory Report, 59GL104, New York: General Electric Company, Schenectady, pp. 227–232.

Tony Adey was already working on a fast model method for two coupled second-order systems when I arrived:

Adey, A. J., Coales, J. F., & Stiles, J. A. (September 1963). 'Predictive control of an on-off system with two control variables'. Proc. 2nd I.F.A.C. Congress, Basle: Switzerland.

My research supervisor was John Flavel Coales, one of the founders of IFAC and its president in 1966. My task was to investigate strategies for third- and higher-order systems:

Billingsley, J. & Coales, J. F. (1968). Simple predictive controller for high-order systems, *Proceedings of the Institution of Electrical Engineers*, Volume 115, Issue 10, October 1968, pp. 1568–1576.

A conclusion we found was that a suboptimal strategy had its own advantages.

The term 'predictive control' was later adopted by other groups and woven into much more complicated strategies, labelled 'Model Predictive Control'. Meanwhile occasional research students worked with me to simplify what now had to be called 'Logical Predictive Control'.

Billingsley, J. & Ghude, S. (2015). Significant advance in logical predictive control, *Electronics Letters*, Volume 51, Issue 16, pp. 1240–1241.

Even more recent developments are simulated in the chapter. Simulation of the ball-and-beam experiment showed how surprisingly great is the improvement with such predictive control.

IN SUMMARY

Please forgive me again for turning this chapter into an ego trip. When I delved back to find the publications that inspired me, it brought a flood of nostalgia. Rather than add a heap of references, these are books and articles that I hope that you might read.

Index

Printed in the United States
by Baker & Taylor Publisher Services